List of titles

Already published

Cell Differentiation	J. M. Ashworth
Biochemical Genetics	R. A. Woods
Functions of Biological Membranes	M. Davies
Cellular Development	D. Garrod
Brain Biochemistry	H. S. Bachelard
Immunochemistry	M. W. Steward
The Selectivity of Drugs	A. Albert
Biomechanics	R. McN. Alexander
Molecular Virology	T. H. Pennington, D. A. Ritchie
Hormone Action	A. Malkinson
Cellular Recognition	M. F. Greaves
Cytogenetics of Man and other Animals	A. McDermott
RNA Biosynthesis	R. H. Burdon
Protein Biosynthesis	A. E. Smith
Biological Energy Conservation	C. Jones
Control of Enzyme Activity	P. Cohen
Metabolic Regulation	R. Denton, C. I. Pogson
Plant Cytogenetics	D. M. Moore
Population Genetics	L. M. Cook
Insect Biochemistry	H. H. Rees
A Biochemical Approach to Nutrition	R. A. Freedland, S. Briggs
Enzyme Kinetics	P. C. Engel
Polysaccharide Shapes	D. A. Rees
Transport phenomena in Plants	D. A. Baker
Cellular Degradative Processes	R. T. Dean
Human Evolution	B. A. Wood

In preparation

The Cell Cycle	S. Shall
Microbial Metabolism	H. Dalton, R. R. Eady
Bacterial Taxonomy	D. Jones, M. Goodfellow
Molecular Evolution	W. Fitch
Metal Ions in Biology	P. M. Harrison, R. Hoare
Muscle	R. M. Simmons
Xenobiotics	D. V. Parke
Human Genetics	J. H. Edwards
Biochemical Systematics	J. B. Harbourne
Biochemical Pharmacology	B. A. Callingham
Biological Oscillations	A. Robertson
Photobiology	K. Poff
Membrane Assembly	J. Haslam

OUTLINE STUDIES IN BIOLOGY

Editor's Foreword

The student of biological science in his final years as an undergraduate and his first years as a graduate is expected to gain some familiarity with current research at the frontiers of his discipline. New research work is published in a perplexing diversity of publications and is inevitably concerned with the minutiae of the subject. The sheer number of research journals and papers also causes confusion and difficulties of assimilation. Review articles usually presuppose a background knowledge of the field and are inevitably rather restricted in scope. There is thus a need for short but authoritative introductions to those areas of modern biological research which are either not dealt with in standard introductory text-books or are not dealt with in sufficient detail to enable the student to go on from them to read scholarly reviews with profit. This series of books is designed to satisfy this need. The authors have been asked to produce a brief outline of their subject assuming that their readers will have read and remembered much of a standard introductory textbook of biology. This outline then sets out to provide by building on this basis, the conceptual framework within which modern research work is progressing and aims to give the reader an indication of the problems, both conceptual and practical, which must be overcome if progress is to be maintained. We hope that students will go on to read the more detailed reviews and articles to which reference is made with a greater insight and understanding of how they fit into the overall scheme of modern research effort and may thus be helped to choose where to make their own contribution to this effort. These books are guidebooks, not textbooks. Modern research pays scant regard for the acedemic divisions into which biological teaching and introductory textbooks must, to a certain extent, be divided. We have thus concentrated in this series on providing guides to those areas which fall between, or which involve, several different academic disciplines. It is here that the gap between the textbook and the research paper is widest and where the need for guidance is greatest. In so doing we hope to have extended or supplemented but not supplanted main texts, and to have given students assistance in seeing how modern biological research is progressing, while at the same time providing a foundation for self help in the achievement of successful examination results.

J. M. Ashworth, Professor of Biology, University of Essex.

Cellular Degradative Processes

R. T. Dean

Research Scientist,
Clinical Research Centre, Harrow

LONDON
CHAPMAN AND HALL

A Halsted Press Book
John Wiley & Sons, New York

First published in 1978
by Chapman and Hall Ltd
11 New Fetter Lane, London EC4P 4EE
© 1978 R. T. Dean
Typeset by Preface Ltd, Salisbury, Wilts
and printed in Great Britain at the
University Printing House, Cambridge

ISBN 0 412 15190 1

Distributed in the U.S.A.
by Halsted Press, a Division of
John Wiley & Sons, Inc., New York

Library of Congress Cataloging in Publication Data

Dean, R. T.

Cellular degradative processes.

1. Cell metabolism. 2. Biodegradation.
3. Lysosomes. I. Title. I. [DNIM: 1. Macromolecular
systems. 2. Cells — Metabolism. 3. Organoids —
Metabolism. QH581 D282c]
QH634.5.D4   574.8'761   77-28244
ISBN 0-470-26300-8

Contents

Preface

This book attempts to give a concise conceptual outline of cellular degradative processes. It covers the specialized degradative organelle, the lysosome, of most eukaryotic cells, and discusses the relationship between breakdown reactions outside the lysosome and those within. Limited degradative reactions (as in the processing of nucleic acids and proteins) are covered, and the emphasis is on breakdown of macromolecules. The turnover of extracellular macromolecules in multicellular organisms is treated, and at the other extreme of the spectrum of biological organization, degradation in bacteria is considered. As is the case with many other important biological processes, macromolecule breakdown seems to occur in a very similar manner in both prokaryotic cells (such as bacteria) and in eukaryotic cells (such as those of man). Thus the book stresses the fundamental similarities of degradative reactions and their central role in the economy of all cells.

R. T. Dean
London, 1977

1 The nature and significance of degradative processes

1.1 Introduction

This book concerns primarily intracellular degradation reactions, which are undergone by most biological molecules. It concentrates on the breakdown of macromolecules (such as the polymeric carbohydrates, nucleic acids and proteins). Most smaller biological molecules undergo several such reactions too (such as oxidation, de-amination, etc) These constitute the classical metabolic pathways, and will not be discussed here in any detail. Instead, the release of such small molecules from the breakdown of the complex cellular macromolecules is considered. Digestion of food in specialized organs of multicellular organisms (such as the alimentary canal of man) is given brief mention, but the degradation of endogenously generated molecules is the central topic of the book.

1.2 Functions and significance of degradative processes

The most familiar set of degradative reactions is digestion, in which extracellular material (food) is converted into small molecules which can be used in the nutrition of the organism. In some primitive organisms (such as Protozoa like *Amoeba* and *Paramecium*) this digestion occurs largely within cells, after uptake of materials by the cells (by the mechanism of endocytosis, which will be discussed in Chapter 5). In higher organisms, with multicellular organization, and considerable differentiation in cell functions, digestion of food is largely extracellular though within the organism. The alimentary canal of man is an example of such a site of extracellular digestion. The products of digestion are taken up by cells for use in biosynthetic reactions.

Some organisms rely upon digestion not only outside their cells, but even outside the bounds of their tissues. For instance bacteria and fungi often secrete digestive enzymes into their environment (which hopefully contains decent amounts of digestible material!), and utilise the products of the ensuing external digestion. Fungi may also use such methods for invading other organisms (such as trees), and the spread of cancer cells through tissues of higher organisms may depend similarly on the secretion of degradative enzymes, which clear a path through the tissues for the cancer cells to traverse. An intermediate form of digestion is exploited by the fascinating insectivorous plants: these have specialized traps in which insects can be immobilized. Once caught, the insect suffers progressive digestion by enzymes which the plant secretes into the extracellular fluid surrounding the insect. The small products of this digestion are then taken up from the trap fluid by the plant and used.

In addition to their extracellular digestive system, the cells of higher organisms have a system for intracellular degradation, the lysosome (see Chapter 4), similar to the internal degradation system of Protozoa. Higher cells exploit the lysosomal system largely in dealing with intracellular materials, but also, particularly in the case of cells active in endocytosis (Chapter 4), in processing external materials taken up by the cells. The intracellular system found in multicellular animal cells is sufficiently like that of unicellular Protozoa for them both to be called lysosomal systems; and although the matter was debated for quite a time, it is now clear that plant cells have a lysosomal apparatus. Only bacteria and viruses lack lysosomes, so the fundamental importance of this organelle is apparent. Bacterial degradative processes are quite similar to those of eukaryotes, in spite of the lack of lysosomes, but bacteria do not endocytose, and thus have to rely on extracellular digestion of food unless they are photosynthetic, and so do not require food at all.

The products of digestion of food, and of intracellular materials, are used as the building blocks for synthesis of macromolecules. Thus amino acids for protein synthesis are provided from breakdown of proteins, nucleic acid precursors from degradation of nucleic acids and so on. Of course the low molecular weight products themselves undergo metabolic conversions, so that carbon skeletons from sugars may be incorporated into fats, and others from amino acids may enter sugars etc. A particularly striking example of the supply of a precursor for synthesis of one family of compounds coming from the degradation of another family has recently come to light.

This is in the synthesis of carnitine, which carries acyl groups within the mitochondrial membrane of many cells, and is important in providing substrate for mitochondrial fatty acid oxidation. The metabolic pathway generating carnitine is shown below. It involves the methylation of protein lysine residues; these may be mono-, di-, or tri-methylated, but only the latter after release from proteins during degradation acts as a precursor of carnitine synthesis (Fig. 1.1)

The necessity of digestion in the economy of organisms which cannot synthesise their own precursor molecules (amino acids etc) is obvious: the external food source has to provide these molecules. This mode of nutrition is described as 'heterotrophic', but, in contrast, some organisms are 'autotrophic' (provide their own metabolites). The main autotrophs are the photosynethic plants and other cells; these use energy from light to drive directly and indirectly the synthesis of all their cellular organic molecules, and only depend on the environment additionally for a supply of inorganic nutrients. Thus ultimately, heterotrophs depend on autotrophs for their supply of food.

But even autotrophs undergo degradative processes similar to those of heterotrophs, although the substrates for degradation are synthesised by the cells. It is not immediately obvious why such cells should require to degrade their macromolecules at all; nor is it obvious why heterotrophs should need to break down their own synthetic products (as opposed to their food, the breakdown of which is clearly necessary).

It must be admitted that we do not understand why rates of degrad-

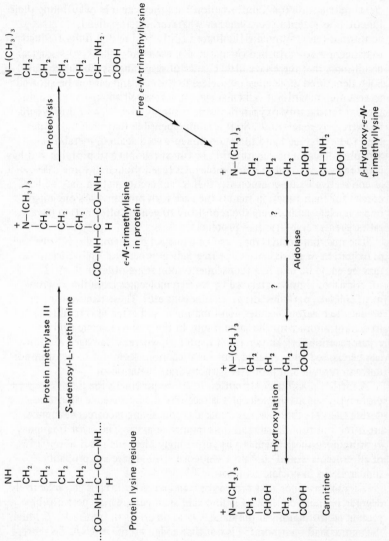

Fig. 1.1 Formation of carnitine from protein-bound trimethyllysine.

ation are as they are; but we can suggest several factors which might make degradation a universal necessity. Foremost amongst these is the limited chemical stability of macromolecules. Thus the polypeptide chains of proteins (the chemistry of which is described in Chapter 2) fold into very complicated three dimensional structures, and only in the so-called native state (or 'Conformation') are they capable of fulfilling their functions as receptors, enzymes or whatever. Macromolecule conformations continually fluctuate slightly, and with a finite frequency they become so perturbed ('denatured'), even under ideal physiological conditions, that they have little chance of returning to the native state. Such denatured molecules are useless to the cell, and cannot be allowed to accumulate without cell enlargement. Gross cell enlargement would prevent satisfactory oxygen or nutrient supply to cells, and thus would be extremely deleterious. So it is comprehensible that such molecules tend to be degraded and their components used again in metabolism. The phenomenon of denaturation has most relevance to proteins, but has some relevance to other macromolecules (even though in some cases, such as connective tissue components their exact conformation may be less crucial for their function than is the case with proteins). As one might hope, nucleic acids, being the repository of genetic information have rather greater stability than proteins.

The macromolecules may also be damaged by chemical reactions such as oxidation or peroxidation, or free radical processes. These reactions may be enzymic in origin (consequent upon some other biological function). or simply catalysed by foreign molecules from the environment (metals, particles, drugs, carcinogens etc). These reactions often render the macromolecules non-functional, and often also tend to lead to denaturation, with the same result. In the case of nucleic acids, some repair reactions are known (see Chapter 5), whereby damaged sections can be excised and replaced, but such reactions seem not to be of importance in protein and probably carbohydrate metabolism.

A further possible contribution to the requirement for the degradation of macromolecules is made by the occurrence of errors during their synthesis. Macromolecules containing incorrect sequences, are often not functional, and thus require removal. The error frequency in protein synthesis seems to be surprisingly high, such that about 15% of all proteins would contain a sequence error. There are probably similar errors in nucleic acid synthesis.

There is evidence that prokaryotic and eukaryotic cells can selectively degrade proteins containing amino acid analogues, which perturb the protein conformation in the same way as do errors in synthesis, in which the amino acid incorporated is not that coded by the mRNA. So there probably is some selective advantage in removing such molecules.

The function of macromolecular carbohydrates and protein-carbohydrate complexes such as glycoproteins seems to depend less critically on the exact sequence of sugars in the carbohydrate chains. However for interactions with receptors on membrane surfaces the exposed terminal residue is often of paramount importance. This may e true of cell—cell interactions mediated by surface glycoproteins, and s certainly true of receptor interactions leading to pinocytosis of circulating glycoproteins by specialised cells (see Chapter 5).

One consequence of these selective pinocytosis systems is that glycoproteins bearing terminal sugars are rapidly removed from circulation and degraded by lysosomes in liver hepatocytes. Other terminal sugars may specify uptake into other cells, which may not necessarily lead rapidly to degradation (for instance phosphohexose residues seem to be important in allowing selective uptake of extracellular lysosomal glycosidases by fibroblasts, and these cells may retain the enzymes for considerable lengths of time). There is little evidence that the internal sequences of sugars in glycoproteins have much structural effect, or any effect on the degradation of the glcoproteins. Of course, these side chains are not synthesised on some coding signal (like an mRNA) but rather by the successive action of discrete glycosyl transferases on extending sugar chains, so that heterogeneity of carbohydrate side-chains on molecules of any single glycoprotein is widespread. Indeed, this is the most general cause of the existence of multiple forms of proteins. This is particularly true of lysosomal enzymes, nearly all of which are glycoproteins, and of blood proteins, most of which are glycoproteins.

It was suggested by Orgel in his 'Error Catastrophe theory' that the progressive accumulation of errors in proteins might be a crucial factor in aging: he pointed out that a protein synthetic machinery comprising proteins containing errors might function progressively more inaccurately. More and more aberrant proteins would be formed, and eventually faulty nucleic acid polymerases would catalyse the formation of faulty nucleic acids, with resultant cell disorganisation and cell death. The presumption that errors would continue to accumulate is not necessarily correct in theory; and observations on the accumulation of error-containing proteins (indicated by abnormal heat lability, or abnormally poor enzyme activity per mole of enzyme protein) have been inconsistent.

The role of degradation in this postulated catastrophe has only been considered recently. The progressive accumulation of errors in either protein or nucleic acid could in theory be prevented by selective degradation of the faulty molecules by means of the processes mentioned already. Conversely, a progressive defect in degradation of such error-containing macromolecules could contribute to an error catastrophe; and such a defect might be expected to follow from the synthesis of degradative enzymes containing incorrect amino acid sequences. Again, experiment has not consistently supported either idea. On balance, it seems that degradative capacity is retained in aging cultured cells, which constitute the main model system. And it seems also that error containing proteins do not generally accumulate because they are rapidly degraded. The increased production of error-containing proteins may be indicated by the observation of an increased proportion of rapidly degraded protein in such aging cells.

The accumulation of peroxidized (damaged) lipid has also been suggested as an important parameter of cellular aging. Again this could reflect an inadequacy in the breakdown of such molecules, as well as an overproduction. This point has not yet received study.

Perhaps the most striking observation amongst those concerned with the degradation of error-containing proteins, is simply the rapidity with which such molecules are removed. The rate may be ten times the average rate of removal of normal proteins. This has led to a more speculative suggestion on the function of turnover of normal macromolecules: that the presence of a turnover system, normally functioning at some basal rates, is required so that when error-containing molecules are produced, they can be removed by accelerated action of this system.

Another theoretical advantage of the existence of basal turnover, is that it allows regulation of macromolecular concentrations; although a rapid increase in the concentration of a macromolecule can theoretically occur simply by a sudden very rapid increase in its rate of synthesis, a rapid decrease requires a rapid rate of degradation. Thus for flexibility in control of macromolecule levels it is desirable to have a degradation system which normally degrades the molecule relatively slowly, but which can do so, when required, at a much greater rate. Indeed, it does seem that many enzymes which regulate rate — limiting steps of metabolic pathways are proteins of rapid turnover, and thus able to respond rapidly to changed requirements. mRNA's for such proteins may well also have rapid turnover, so that after a temporary increase in level allowing rapid synthesis of the protein, the mRNA can be degraded rapidly to bring it back to its normal concentration.

Degradative processes also provide protective functions; for instance, multicellular organisms combat bacterial and viral infection by degrading the invading organisms. Many are endocytosed by 'professional' phagocytic cells (see Chapter 4) circulating in blood and extracellular fluids, and degraded intracellularly. Foreign chemicals other than foodstuffs may meet a similar fate; the products of digestion of such molecules may be excreted or used according to their nature, while this may not be possible with the intact foreign molecule, perhaps because it is too large in molecular size.

Endocytosis followed by degradation may also be important in the termination of certain signals to cells provided at the cell membrane (by hormones, immune stimuli etc). In these cases the surface stimulus is transduced into a cellular activity which usually includes the removal of the stimulus from the cell surface by endocytosis and degradation. The stimuli and their transduction may themselves involve degradation of macromolecules (see Chapter 6). In addition some hormones seem to be degraded at the cell surface, possibly by a non-specific mechanism capable of acting on several hormones.

There are also circumstances in which extracellular degradation both of whole cells and of macromolecules (other than food) is important. Examples occur in the early stages of killing of invading organisms, (which is usually commenced outside phagocytic cells by means of surface enzymes on the phagocytes, and by secretion of enzymes such as lysosomal hydrolases by the phagocyte), and in processes of tissue remodelling, or of tissue regression in development. An example of remodelling is the conversion of zones of cartilage into bone, a stronger connective tissue. Most of the material degraded is extracellular, but in contrast, tissue regression in development often requires digestion of whole cells (see Chapter 5). For instance, in the development of human males, the Mullerian duct regresses; while in tadpoles, the tail is degraded. In these cases degradation precedes a necessary constructive phase.

Directly useful examples of intracellular degradation have been appreciated recently. Many macromolecules seem to be synthesized in the form of precursors which are even larger than the normal functional molecule. This is true of nucleic acids, where large areas of nucleotide sequence which apparently do not code for functional amino acid sequences, and which are often not transcribed (from DNA to RNA) or translated (from RNA to protein), have been discovered. Such areas may be removed from RNA molecules in the nucleus, before the RNA is transported to the cytosol, while other areas of this kind may have functions to be carried out in the cytosol (this is discussed further in Chapter 5).

Similarly, many proteins are made as larger molecular weight precursors; in the case of enzymes these tend to be relatively inactive, and to

Table 1.1 Physiological systems involving limited proteolysis

Physiological system	Example
Defence reactions	Coagulation Fibrinolysis Complement activation
Hormone production	Proinsulin → Insulin Angiotensinogen → Angiotensin Proparathyroid hormone → Parathyroid hormone Proglucagon → Glucagon Large Gastrin → Gastrin
Assembly	Many viral proteins Procollagen → Collagen Proalbumin → Albumin Fibrinogen → Fibrin Some intracellular proteins, such as the peroxisomal enzyme, catalase.
Development	Activation of prochitin synthetase in the initiation of septum formation in budding yeast Activation of procoocoonase used in the escape of certain silk moths from their cocoons. Fertilization (proacrosin-acrosin)
Tissue injury	Prekallikrein → kallikrein Kininogen → kinin under the action of kallikrein. Kinins are hypotensive agents with significant inflammatory effects.

become activated when the precursor sequence of polypeptide is removed by proteolysis. At this stage, presumably some refolding occurs. Ironically, the first known examples were the pancreatic digestive enzymes such as trypsin, which are secreted as inactive forms ('zymogens') which become active after proteolytic cleavage of their polypeptides; in some cases this seems to be autocatalytic, in that even the zymogen has finite proteolytic activity against itself, and the resultant active form may then catalyse the activation of its own, or another zymogen.

Subsequently, it was shown by Steiner, that the protein hormone insulin is also synthesised as a precursor ('proinsulin') but converted to insulin before secretion. Since then, the production of pro-forms which undergo limited cleavage to produce the native molecule, has been shown to apply to several other secretory proteins, such as albumin, the main blood protein in higher animals. The possible mechanism of this process is described in Chapter 4.2. Several intracellular proteins seem also to be synthesized via pro-forms. A partial list of molecules known to be synthesized as precursors requiring proteolytic activation is given in Table 1.1.

'Pre-pro' forms of secretory and other proteins have also been discovered recently. The 'pre' designation refers to a second sequence of polypeptide present only in precursor forms and removed before the molecule becomes functional. Known presegments are of uniform length (about twenty amino acids) consistent with the suggestion that they are responsible for transferring the polypeptide across the endoplasmic reticulum membrane into the lumen. Such a transmembrane transport mechanism is essential for the secretion of such macromolecules since they cannot freely penetrate membranes.

Extracellular proteolytic activation is involved in three cascades systems of proteinases present in blood: the complement system, the coagulation system, and the fibrinolytic system. The first is concerned with protection against foreign materials (antigens) which have caused an immune response, though in some cases it may be activated directly (by foreign particles and compounds, or by cell-derived proteinases): activated complement components have diverse effects, including lysis of membranes, and stimulation of endocytosis and secretion. The coagulation system is responsible for the formation of blood clots of fibrin, which are important in wound healing; while fibrinolysis is required during the recovery phase after coagulation. These mechanisms are discussed further in Section 5.2.

Bibliography

Brooks, F. P., (1970), *Control of Gastrointestinal Function*, Macmillan, London. *A basic review of digestion and absorption in man.*
DePierre, J. W. and L. Ernster, (1977), Enzyme topology of intracellular membranes, *Ann. Rev. Biochem.*, 46, 201–262.

Covers transfer of proteins across the endoplasmic reticulum membrane, together with other aspects of topology of biosynthesis of proteins.

Heslop-Harrison, Y., (1975), Enzyme Release in Carnivorous Plants, *In Lysosomes in Biology and Pathology*, Vol. 4, (Eds.) J. T. Dingle and R. T. Dean, pp. 525–578, North Holland, Amsterdam.
A clear review with some beautiful photographs.
Hughes, R. C., (1977), Recognition of lysosomal enzymes, *Nature* (London), 269–270.
A short review of specific receptors for terminal sugars on glycoproteins.
Matile, H., (1975), *The Lytic Compartment of Plant Cells*, Springer, Vienna
Includes a discussion of secretion enzymes by fungi, together with the generalities of plant lysosomes.
Orgel, L. E. (1963), *Proc. natn. Acad. Sci.,* U.S.A., **49**, 517–521.
Orgel, L. E. (1973), Nature (London), **243**, 441–445
On the 'Error Catastrophe' Hypothesis.
Paik, W. K., Nachunson, S., and Kim, S., (1977), Carnitine biosynthesis via protein methylation, *Trends in Biochemistry Science*, **2**, 159–161.
Rothman, J. E. and Lodish, H. F., (1977), *Nature (London), 269*, 775–780.
Role of pre(signal) sequences in proteins.
Steiner, D. F., Kemmler, W., Tager, H. S., Rubenstein, A. H., Lernmark, A. and Zuhlke, H., (1975), Proteolytic mechanisms in the biosynthesis of poly-peptide hormones, *in Proteases and Biological Control*, (Eds.) E. Reich, D. B. Rifkin and E. Shaw, pp. 531–549, Cold Spring Harbour.
Good review.

2 Methods and problems in the study of turnover; a very brief survey

The initial aim of the study of turnover is somehow to 'label' the population of molecules of a particular substance (an enzyme maybe) or group of substances (proteins in general perhaps) present at a particular instant in order to follow the survival of that same set of molecules as time proceeds. The degradation of the molecule or class of molecule usually follows exponential kinetics, and can conveniently be described in terms of a 'half-life' which is the time for half of the molecules originally present to be degraded (and in the steady state, replaced).

The normal means of labelling is to incorporate radioactive precursors of the molecules of interest (amino acids for proteins) by biosynthetic means (rather than by artificial chemical reactions which might them-selves affect the cell behaviour). The choice of precursor is of particu-lar importance for several reasons. Ideally it should only be incorporated into the molecules of interest, so that the measurement of radioactivity unambiguously reflects that component. With the complexity of metabolic pathways, this is quite difficult to achieve. But this may not be crucial as purification procedures will have to be employed to isolate the molecules under study, even if they are a whole class like the proteins.

When the turnover of a single protein (or other macromolecule, such as the mRNA for a particular protein) is under investigation, complete purification of that protein is normally required; and for fully reliable information about its turnover, it is necessary that at each time studied, radiochemical homogeneity of the protein is ensured. This means that a demonstrable maximum ratio of radioactivity to molecules of protein should be reached by at least two different methods. One can easily see that this involves a massive amount of work: and it is all too often neglected.

The other requirement which is crucial is that the precursor should be removed completely after its initial incorporation, so that the time of commencement of decay of radioactivity in the molecule under study is known unambiguously. A related requirement is that radioactive molecules released from degradation of the macromolecule should not be reincorporated into further molecules ('re-utilized'). This problem was not appreciated in early work, and led to vast underestimates of the rate of degradation of proteins. Recent values have tended to be bigger, because precursors which are rapidly degraded themselves into molecules which are no longer precursors have been used. Alternatively, re-utilization can be suppressed in many cases by the presence during the degradation period of a very high concentration of non-radioactive molecules of the precursor. These molecules will contribute the vast majority of precursor molecules being incorporated into new macromolecules, and thus re-utilization is suppressed. Problems of suppressing re-utilization in the study of nucleic acid turnover are only now under attack. In the case of lipids, the choice of precursor is particularly difficult, as degradation tends to lead to interconversion of lipids. So in this case it is particularly important to show that the radioactivity is in the molecule of interest and not just in a closely similar one.

As mentioned already phospholipid turnover is complicated by the exchange of lipids between cell organelles, and between the outside and inside of cells. But of course exchanges of protein molecules between organelles also occur (for instance after synthesis during transport to the functional site, such as within mitochondria). Thus loss of radioactive protein from a cell organelle may reflect degradation or transport and these two should always be clearly distinguished experimentally.

Measurements of the absolute number of molecules of a particular component made or degraded in a given time introduce fresh complications. This is particularly so when, for instance, bulk protein turnover is under investigation. Since it is extremely difficult to relate the radioactivity in protein to the total number of protein molecules, or to the total number of amino acids in protein, a converse approach is often used. This is to measure the rate of synthesis of protein, to measure the time dependence of the amount of total protein in the system, and to deduce protein degradation by difference. To relate incorporation of radioactive amino acids to incorporation of molecules of that amino acid, it is necessary to know how much radioactivity per molecule of that amino acid there is in the actual pool of amino acids from which protein synthesis takes place (and not merely in the starting amino acid mixture added). This has proved extremely difficult, particularly as the identification of such a pool has not unambiguously been made. One approach which seems to be valuable, is to supply radioactive precursor continuously, until all pools of precursor reach constant, and hopefully similar, ratios of radioactivity to molecules of precursor. When this state obtains, the ratio can be determined simply from material in the extracellular medium which of course has terrific practical advantages.

Many other methods have been used in the study of turnover besides those which fall into the categories outlined above. But many are subject to the same complexities, plus extra ones induced by the use of inhibitors (of protein synthesis, for instance) which have many direct and indirect effects. Many of the treatments perturb an ongoing steady state: and of course it is diffuclt to get meaningful measurements in the face of a changing situation. Altogether, the problems are formidable, and the methods more and more sophisticated. But it is hoped that the succeeding chapters will at least convince the reader of the rewards to be obtained from overcoming these difficulties!

Bibliography
Golberg, A. L. and Dice, J. F., (1974), *Ann. Rev. Biochem.*, 43, 835–869. *Methodology.*

3 The chemistry of biological macromolecules and their degradation

The nature of each of the main groups of cellular macromolecules is outlined here, and followed in each case by a description of the main degradative enzymes likely to act on the substrates, and their mode of action. The most important degradative reaction is hydrolysis; to illustrate the immense potential of cellular hydrolytic enzymes, the current classification of these enzymes, together with their enzyme group numbers (E.C.) is shown in Table 3.1. Since each group comprises several enzymes, the enzyme armoury is impressive.

3.1 The structure of nucleic acids
That of DNA is illustrated in Fig. 3.1. DNA is usually in the form of two anti-parallel strands, in which complementary base pairs hydrogen bond

Table 3.1 The classification of hydrolases, according to the enzyme commission, with E.C. numbers.

Enzyme class	Subclass

3 HYDROLASES

3.1	*Acting on ester bonds*
	3.1.1 Carboxylic ester hydrolases
	3.1.2 Thiolester hydrolases
	3.1.3 Phosphoric monoester hydrolases
	3.1.4 Phosphoric diester hydrolases
	3.1.5 Triphosphoric monoester hydrolases
	3.1.6 Sulphuric ester hydrolases
	3.1.7 Diphosphoric monoester hydrolases
3.2	*Acting on glycosyl compounds*
	3.2.1 Hydrolysing O-glycosyl compounds
	3.2.2 Hydrolysing N-glycosyl compounds
	3.2.3 Hydrolysing S-glycosyl compounds
3.3	*Acting on ether bonds*
	3.3.1 Thioester hydrolases
	3.3.2 Ether hydrolases
3.4	*Acting on peptide bonds (peptide hydrolases)*
	3.4.11 α-Aminoacylpeptide hydrolases
	3.4.12 Peptidylamino-acid or acylamino-acid hydrolases
	3.4.13 Dipeptide hydrolases
	3.4.14 Dipeptidylpeptide hydrolases
	3.4.15 Peptidyldipeptide hydrolases
	3.4.21 Serine proteinases
	3.4.22 SH-proteinases
	3.4.23 Acid proteinases
	3.4.24 Metalloproteinases
	3.4.25 Proteinases of unknown catalytic mechanism
3.5	*Acting on carbon−nitrogen bonds, other than peptide bonds*
	3.5.1 In linear amides
	3.5.2 In cyclic amides
	3.5.3 In linear amidines
	3.5.4 In cyclic amidines
	3.5.5 In nitriles
	3.5.99 In other compounds
3.6	*Acting on acid anhydrides*
	3.6.1 In phosphoryl-containing anhydrides
	3.6.2 In sulphonyl-containing anhydrides
3.7	*Acting on carbon−carbon bonds*
	3.7.1 In ketonic substances
3.8	*Acting on halide bonds*
	3.8.1 In C−halide compounds
	3.8.2 In P−halide compounds
3.9	*Acting on phosphorus−nitrogen bonds*
3.10	*Acting on sulphur−nitrogen bonds*
3.11	*Acting on carbon−phosphorus bonds*

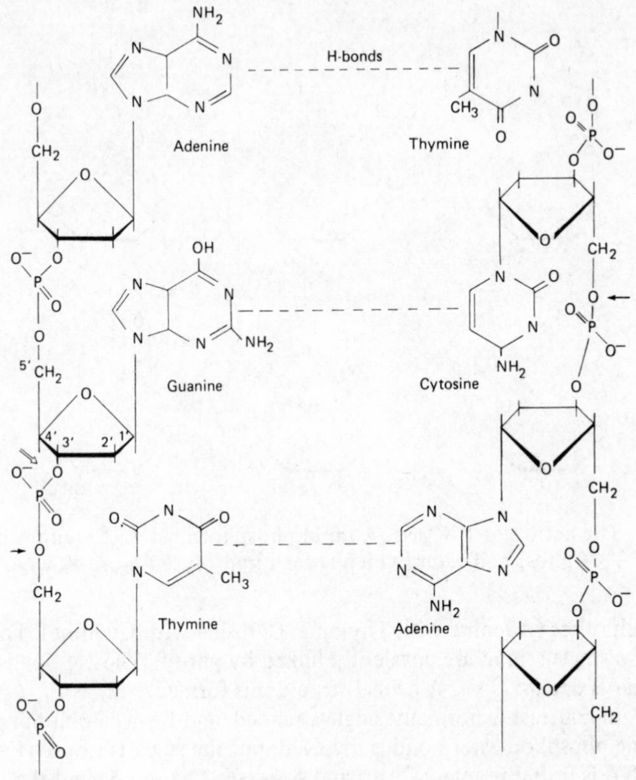

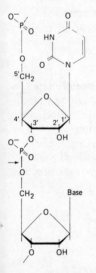

Fig. 3.1 The structure of DNA. Arrows indicate the sites of attack at nucleases to be discussed later (Section 3.2). → DNAase II ⇒ DNAse I

Fig. 3.2 The structure of RNA. The arrow → indicates the site of attack of RNAse (Section 3.2).

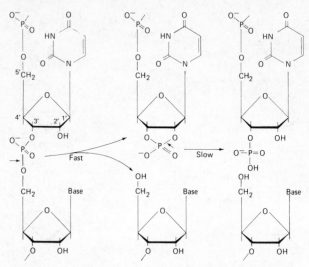

Fig. 3.3 The action of RNAase. A rapid phosphotransterase step produces a cyclic 2',3'-phosphodiester, which is then hydrolysed more slowly.

with each other (Adenine with Thymine; Cytosine with Guanine). The units of a single strand are covalently linked by phosphodiester bonds. The sugar is deoxyribose. A helical structure is formed.

RNA in contrast is normally single stranded, and forms a random coil. The same phosphodiester bond is involved, but the sugar is ribose (Fig. 3.2). RNA is found in many forms and sizes (see Chapter 5) and the smallest group, the transfer RNA (tRNA) has more specialized conformation, connected with their function of transfer of amino acids.

Most nucleic acid occurs in close association with proteins; and the histone and acid proteins associated with DNA may well be concerned in regulating its expression. It is not clear whether degradation or other modifications of the proteins are involved in this.

3.2 Nucleases

There are two main types of DNAses. DNAse 1, which has been isolated from pancreas, shows a neutral pH optimum, and acts on only one strand, to liberate a 5'-phosphate terminus. DNAse II is a lysosomal enzyme, normally isolated from spleen, and shows an acid pH optimum. It performs rapid two strand attack, liberating 3'-phosphate termini. The sites of action of both enzymes are shown on Fig. 3.1. As with all the other enzymes to be discussed in any detail, the mechanism of scission of the phosphodiester bond is a hydrolytic one:

$$\text{Nucleoside—3'—O—} \underset{\underset{\text{Nucleoside—5'—O}}{|}}{\overset{\overset{\text{O—}}{\parallel}}{P}} =O + H_2O \rightleftharpoons \text{Nucleside—3'—O—} \underset{\underset{\text{OH}}{|}}{\overset{\overset{\text{O—}}{\parallel}}{P}} =O$$

$$+ \text{ Nucleoside—5'—OH}$$

20

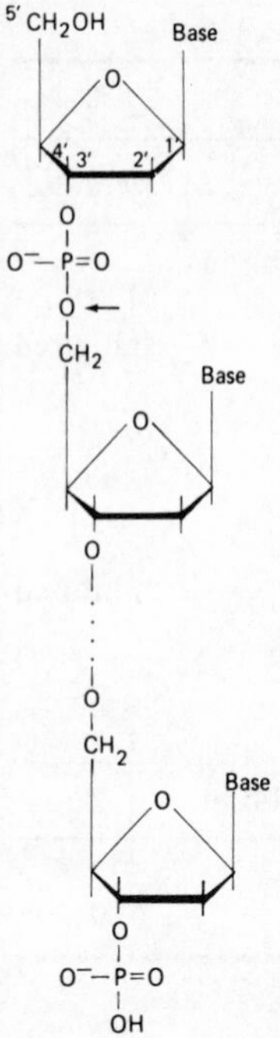

Fig. 3.4 The site of attack of phosphodiesterase II on a fragment of DNA.

There are also two main types of RNAse: an enzyme (usually obtained from pancreas) with an alkaline pH optimum, and the lysosomal RNAse II, with an acid pH optimum. The mode of action of both enzymes seems to be similar: a phosphate group from the $5'$-position of one nucleotide is transferred to the $2'$-position of the next, forming a $2', 3'$-cyclic phosphate diester, which is then slowly hydrolysed. The lysosomal enzyme depolymerizes soluble RNA (of small molecular weight compared with ribosomal RNA) nearly as well as yeast ribosomal RNA. It acts on both purine and pyrimidine nucleotides. The reaction pattern of the RNAses is illustrated in Fig. 3.3.

The oligonucleotides released from the initial breakdown of DNA and RNA can be degraded to mononucleotides under the action of several enzymes. Chief among these are phosphodiesterase I, which has an alkaline pH optimum and removes $5'$-phosphomononucleotides from the $3'$-hydroxyl end of oligonucleotides, and phosphodiesterase II. The latter is a lysosomal enzyme, with an acid pH optimum. Its action to liberate $3'$-phosphomononucleotides from the $5'$-hydroxyl end of oligonucleotides, is shown in Fig. 3.4.

3.3 Proteins
The general structure of proteins is:

$$NH_2 R^1 - CO - NH - R^2 - CO - NH \cdots R^n\ COOH$$

where $R^1 \cdots R^n$ are amino acids, the arrangement being characteristic of

21

Table 3.2 Criteria for the classification of endopeptidases

Inhibitor or activator (to be used with 1 h preincubation at 20°C)	Class of endopeptidase			
	Serine	Thiol	Carboxyl	Metallo
Pepstatin (1 µg/ml)			Inhibited	–
Dithiothreitol + EDTA (each 2 mM)	–	Activated	–	Inhibited
Dip-F or Pms-F (1 mM)	Inhibited	(Inhibited)	–	
Lima bean or soya bean trypsin inhibitor (100 µg/ml)	Inhibited		–	
1,10-Phenanthroline (1 mM)	–	–	–	Inhibited
4-Chloromercuribenzoate (1 mM)	–	Inhibited	–	–
Diazoacetylnorleucine methyl ester + Cu^{2+} (each 1 mM)	–	Inhibited	Inhibited	–
pH Optimum expected in the range:	7–9	4–7	2–5	7–9

the protein. The bond between adjacent amino acids is called a peptide bond, and it is hydrolysed by proteinases as follows:

$$R^1-CO-NH-R^2 + H_2O \; \rightleftharpoons \; R^1COOH + R^2NH_2$$

3.4 Proteinases

Proteinases, in the general sense of enzymes which degrade proteins can be divided into exo- and endo-peptidases. The former act near the ends of polypeptide chains, while the latter act within the chains. The exopeptidases can be classified on the basis of their specificity (groups 3.4.11 to 3.4.15 in Table 3.1), but the endopeptidases have much too complex specificities for this to be possible. Instead they can most usefully be classified on the basis of their catalytic mechanism. The identity of the essential catalytic group in the enzyme is used to designate the four classes. A fifth class is used for those which presently do not fit in simply. Some of the criteria which can be used to assign an endopeptidase to its class are shown in Table 3.2.

Exopeptidases are well represented both in cytosol and lysosomes of cells, but intracellular endopeptidases are rather concentrated in lysosomes where they may be at very high levels (in liver lysosomes Cathepsins D and B seem each to be about 1 mM!). There are several known proteinases, usually active at neutral pH, which have been found in Cytosol, mitochondria and ribosomes of cells. Knowledge about proteinases in other parts of the cell is more restricted.

Table 3.3 Structure and occurrence of polysaccharides

Oligo- or polysaccharide	Source	Acceptor or repeating unit	Linkages
Sucrose (-P)	Plants	D-Fructose (-6-P)	α, 1—2
Lactose (-P)	Mammals	D Glucose (-1-P)	β, 1—4
Sialyl-lactose	Mammals	Lactose	α, 1—3 (galactose)
Fucosyl-lactose	Mammals	Lactose	α, 1—3 (galactose)
α-1,4-Glucan starch amylose	Plants	D-Glucose	α, 1—4
α-1,4-Glucan glycogen amylose	Mammalian liver	D Glucose	α, 1—4
β-1,3-Glucan callose)	Beans	D Glucose	β, 1—3
β-1,4-Glucan cellulose)	Mung beans, bacteria	D-Glucose	β, 1—4
β-1,4-Xylan	Plants	D-Xylose	β, 1—4
chitin	*Neurospora crassa*	N-Acetyl-D-glucosamine	α, 1—4
colominic acid	*E. coli*	N-Acetylneuraminic acid	α, 1—5
Hyaluronic acid	Bacteria, animals (umbilicilicus, vitreous humour, etc.)	D-Glucuronic acid; N-acetylglucosamine	β, 1—3 β, 1—4
Chondroitin	Mammalian cornea	D-Glucuronic acid; N-acetyl-galactosamine	β, 1—3 β, 1—4
Chondroitin sulphates	Mammalian connective	D-Glucuronic acid; N-acetyl-D-galactosamine-4-, or -6-sulphate	β, 1—3
Dermatan sulphate	Mammalian skin	L-Iduronic acid; N-acetyl-D-galactosamine-2-sulphate	α, 1—3
Heparin	Mammalian liver, lung arterial walls	D-Glucuronic acid 2-sulphate; D-galactosamine-N,C-6-disulphate	α, 1—4 α, 1—4

Table 3.3 (contd.)

Oligo- or polysaccharide	Source	Acceptor or repeating unit	Linkages
Capsular polysaccharide	Type III pneumocci	D-Glucuronic acid; D-glucose	β, 1–4 β, 1–3
Capsular polysaccharide	Type VIII pneumococci	D-Glucuronic acid D-Glucose D-Galactose	
Teichoic acids	Bacteria	Polyribitol or polyglycerol-*P* with D-glucose or *N*-acetyl-glucosamine in glycosidic linkage	α, 1–2
Murein	Bacteria cell walls	*N*-Acetyl-D-glucosamine *N*-Acetylmuropeptides	β, 1–6 β, 1–4
Glycoproteins	Mammalian submaxillary gland	*N*-Acetylneuraminic acid *N*-Acetyl-D-galactosamine disaccharide	α, 2–6
Glycoproteins	Mammalian plasma: fetuin, orosmomucoid	*N*-Acetylneuraminic acid *N*-Acetyl-D-glucosamine D-Mannose	α, 2–6
Blood group substances	Animals	L-Fucose, *N*-acetyl-D-galacto-samine D-galactose	

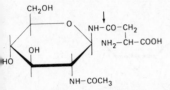

Fig. 3.5 The action of β-aspartyl-glucosylaminase.

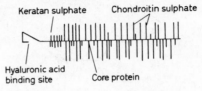

Fig. 3.6 The cartilage proteoglycan molecule.

3.5 Carbohydrates and glyco-conjuagates: structure

Many different polysaccharides are known. They may comprise a repeating unit of one or more sugars, or may have less regular repetitions of fixed monomeric constituents. Some of the main kinds are listed in Table 3.3, together with a little information about the components of the carbohydrate side chains of glycoproteins.

Amongst the polysaccharides are many storage compounds such as glycogen (in mammalian liver for example) and starch (in plants). The degradation routes of these materials have been relatively well studied, and the control of such reactions is outlined in Chapter 5.3 and 6.2.

One of the main chemical linkages between the carbohydrate moitety and the polypeptide of glycoproteins is that between the Cl of N-acetyl-glucosamine and the amide nitrogen of asparagine, shown in Fig. 3.5. Thus in degrading glycoproteins, amidase action is required as well as glycosidase and proteinase action.

While the glycoproteins comprise a large protein moiety with a relatively small amount of carbohydrate, the converse is true of the connective tissue proteoglycans which are conjugates of protein with glyco-aminoglycans such as chondroitin sulphate, keratan sulphate etc (see Table 3.3). Aggregation is an important factor in the structure of proteoglycans and their chemistry is still not fully understood. Fig. 3.6 gives an outline of their molecular arrangement. Fig. 4.7 indicates the degradation of chondroitin sulphate by the concerted action of several lysosomal glycosidases (see Chapter 4).

3.6 Glycosidases

A wide array of glycosidases are known. In bacteria and plants, and in the cytosol of higher cells there are several glycosidases with neutral pH optima, such as the amylases, which can degrade starch and glycogen. But in mammalian cells, the best known are the lysosomal acid glycosidases. However the best studied of all is hen egg white lysozyme on which detailed structural work has been performed. It hydrolyses the (1β-4)-linkages of N-acetylmuramic acid to N-acetylglucosamine in some bacterial wall polysaccharides. It is also found in phagocytic cells, where it aids in the killing and degradation of invading bacteria. Its reaction is shown below (Fig. 3.7).

Lysozyme is present in the lysosomes of some phagocytes, but it is not a typical lysosomal enzyme, and is absent from most cells. In contrast the remainder of the diverse array of lysosomal glycosidases are

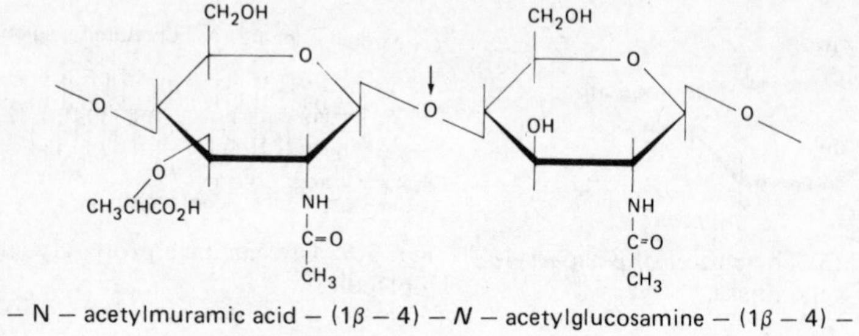

— N — acetylmuramic acid — (1β – 4) — **N** — acetylglucosamine — (1β – 4) —

Fig. 3.7 The repeating disaccharide unit from the cell wall of *Micrococcus lutcus*, showing the point of attack by lysozyme.

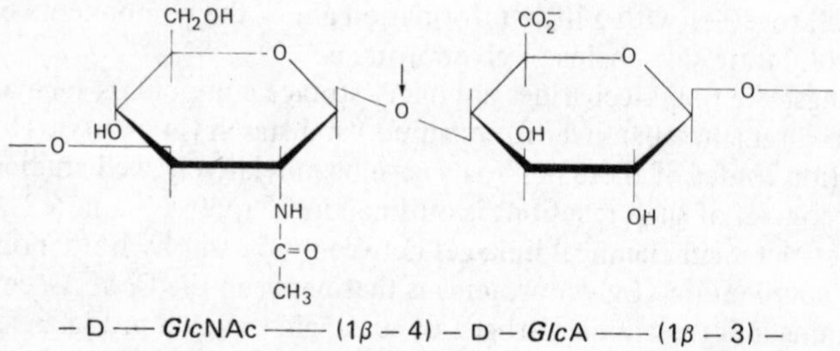

— D —— **Glc**NAc —— (1β – 4) – D – **Glc**A —— (1β – 3) –

Fig. 3.8 The site of attack of hyaluronidase on hyaluronic acid.

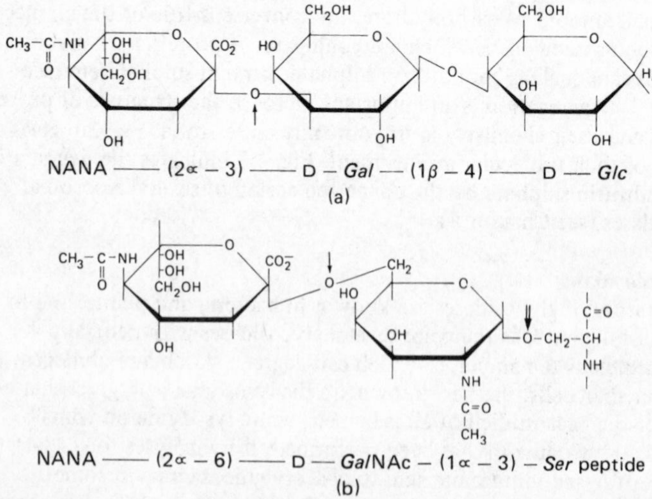

NANA —— (2α – 3) —— D – **Gal** — (1β – 4) —— D —— **Glc**
(a)

NANA —— (2α – 6) —— D – **Gal**NAc – (1α – 3) – *Ser* peptide
(b)

Fig. 3.9 (a) Sialyllactose, showing the point of action of neuraminidase from lysosomes; (b) The glycopeptide of ovine submaxillary mucin, linked via serine to the polypeptide, showing the site for neuraminidase (→), and a subsequent site for β-N-acetylgalactosaminidase (⇒).

26

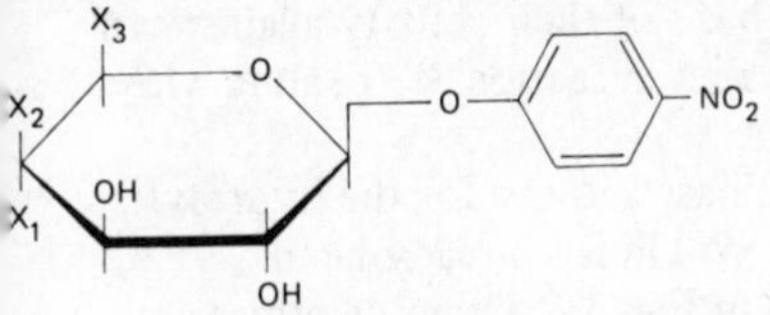

	X_1	X_2	X_3
β – D – glucoside	OH	H	CH_2OH
β – D – xyloside	OH	H	H
β – D – galactoside	H	OH	CH_2OH
β – D – fucoside	H	OH	CH_3
α – L – arabinoside	H	OH	H

Fig. 3.10 Some p-nitrophenyl substrates for glycosidases

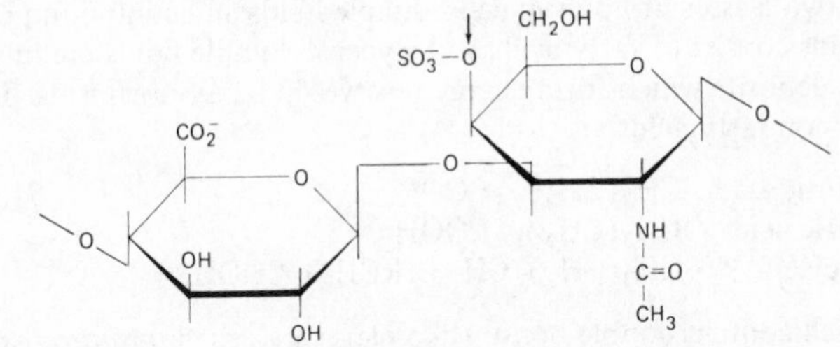

– D —— GlcA —— (1β – 3) —— D — GalNAc – 4SO₄ – (1β – 4) –

Fig. 3.11 The site of attack of chondroitin-4-sulphate by sulphatase B.

present in most cells. They are capable of dealing with most biological carbohydrates. For instance there are several glycosidases which share with lysozyme the capacity for cleaving bonds within polysaccharide chains (and thus are termed endoglycosidases). Fig. 3.8 illustrates the action of hyaluronate endoglucosaminidase (common name hyaluronidase) of the repeating disaccharide of hyaluronid acid. Endoglycosidases are known which act on heparin, glycopeptides (in glycoproteins) and heparan sulphate and many others are suspected.

Even more exoglycosidases have been characterized. For instance the action of neuraminidase in liberating neuraminic acid from several compounds is shown in Fig. 3.9. The enzyme also acts on glycolipids (see Section 3.7). Exoglycosides are conveniently assayed and studied by means of their action on chromogenic or fluorogenic substrates. Some of the usual substrates are illustrated in Fig. 3.10. The liberated nitrophenol is detected as a yellow colour at alkaline pH, while the umbelliferone is fluorescent in that range. In the case of lysosomal enzymes, of course, the assay incubations take place at acid pH, and then the products are measured at alkaline pH.

One of the more remarkable glycosidases is the cytosol β-glucosidase of liver. This attacks a very wide array of such synthetic substrates (all those in Fig. 3.10), whereas most glycosidases only attack one such substrate. However, much more important is their capacity for degrading biological macromolecules. This has only recently received much attention, but has already allowed differentiation between many enzymes which were not clearly distinct on the basis of their activity against synthetic substrates. One of these is illustrated in section 3.7 dealing with glycolipids.

The necessity of aspartylglucosylaminase activity for the hydrolysis of the link between carbohydrate and protein has already been mentioned; its site of activity is shown in Fig. 3.5. Other enzymes are known which can deal with the other sugar—protein links which have been found. In addition many sulphatases are able to remove the sulphate groups from the sulphated glycosaminoglycans. The action of one such enzyme is shown in Fig. 3.11

3.7 Lipids
There are two classes of lipids in cells: simple lipids and compound lipids. Simple lipids consist of fatty acids and glycerol. Simple lipids are important in fat deposits which form energy reserves in many organisms. The most common fatty acids are:

Palmitic acid $CH_3(CH_2)_{14}COOH$
Stearic acid $CH_3(CH_2)_{17}COOH$
Oleic acid $CH_3(CH_2)_7CH=CH(CH_2)_7COOH.$

Those which contain double bonds (like oleic) are called *unsaturated*, while those without are termed *Saturated*.

Compound lipids are more complex, and in addition to fatty acids, may include glycerol and similar compounds, and nitrogenous bases.

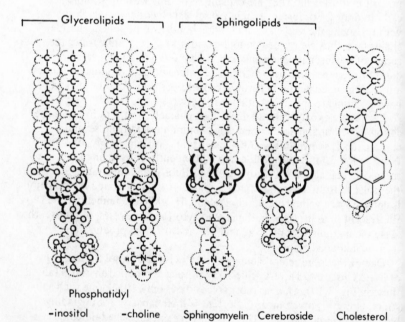

Phosphatidyl

-inositol   -choline   Sphingomyelin   Cerebroside   Cholesterol

Fig. 3.12 Major lipids. (Material from Finean, J. B. (1961), in *Chemical Ultrastructure of Living Tissue*, C. C. Thomas, Illinois.)

Steroids do not contain fatty acids but are often classified with this group. The most important categories of compound lipids are glycerophosphatides, sphingolipids and steroids.

Glycerophosphatides are based on simple lipids but one of the fatty acids is substituted. For instance, amongst this group are phosphatidyl choline (lecithin) which comprises glycerol, two fatty acids, phosphate and the nitrogenous base, choline. Phosphatidyl inositol is similar but choline is replaced by inositol. These compounds are often referred to as 'phospholipids'.

Sphingolipids, the second class of compound lipids, lack glycerol, unlike the glycerophosphatides. They contain instead the nitrogenous base sphingosine. Amongst this class are sphingomyelin (sphingosine, one fatty acid, phosphate and choline) and cerebroside (sphingosine, one fatty acid and galactose). The latter compound can also be described as a glycolipid, since it contains a sugar.

Steroids, the third class of compound lipid, include the major membrane component, cholesterol.

The structures of some of the major lipids are shown in Fig. 3.12 which indicates their approximate spatial disposition.

3.8 Lipases

Many such enzymes are found both in lysosomes and on cellular membranes, and in the soluble phase of cells. They may be released extracellularly on occasion also (besides, of course, in the gut). Thus again enzymes of both acid and neutral or alkaline pH optima are known. Fig. 3.13 indicates the points of attack of four phospholipases on phosphatidylcholine.

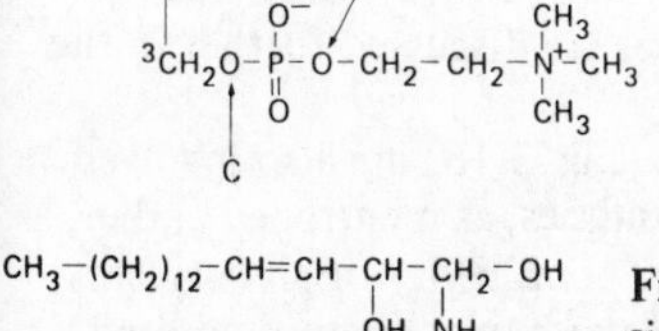

Fig. 3.13 The point of attack of phospholipases A_1, A_2, C and D on phosphatidylcholine.

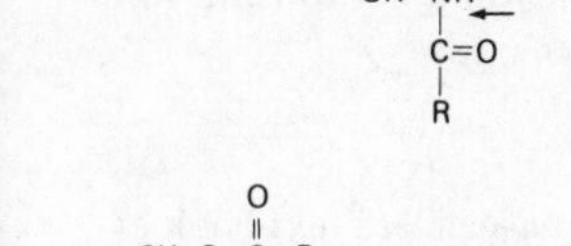

Fig. 3.14 A ceramide (acylsphingosine), showing the point of attack by acylsphingosine deacylase. R = aliphatic chain of fatty acyl radical.

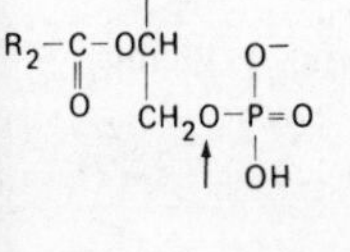

Fig. 3.15 The point of attack of phosphatidate phosphatase on a phosphatidate. R_1, R_2 represent aliphatic chains of fatty acyl radicals.

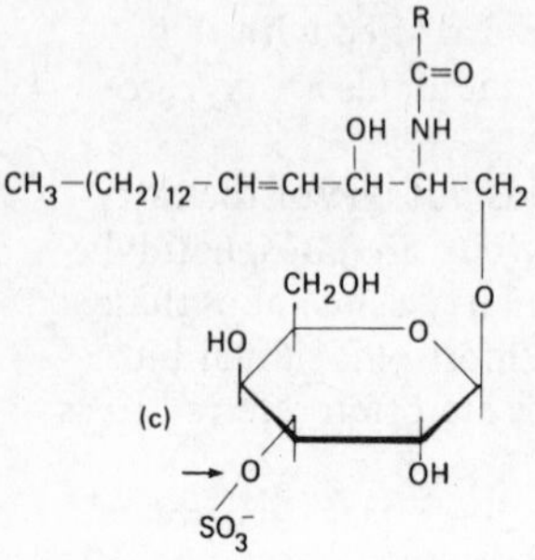

Fig. 3.16 The action of sulphatase A on cerebroside-3-sulphate.

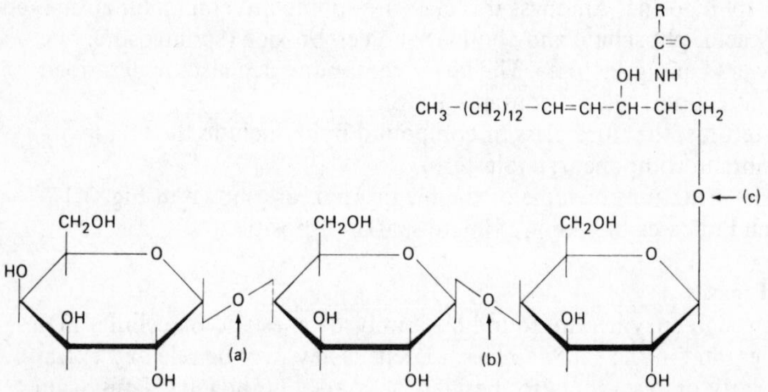

Fig. 3.17 Ceramide trihexoside (galactosyl-β-galactosyl-β-glucosyl-ceramide). Arrows indicate glycosidase attack by: (a) α-galactosidase, and successively; (b) β-galactosidase; (c) β-glucosidase. R- represents an aliphatic chain of the fatty acyl radical.

The characteristics of several lysosomal enzymes acting on lipids have become better understood as a result of work on the lysosomal storage diseases, in which deficiency of single enzymes leads to lethal storage of materials, which would normally be successfully degraded. Fig. 3.14 illustrates the action of an enzyme, acylsphingosine deacylase, which is deficient in Farber's disease, in which its normal substrate accumulates within lysosomes. Lysosomal storage diseases are discussed further in the following Chapter (4).

Phosphatases (Fig. 3.15) and sulphatases (Fig. 3.16) are also involved in the catabolism of lipids. So too are glycosidases, as mentioned earlier, and the storage diseases have revealed a great complexity of glycosidases acting on glycolipids. The action of some of these on a compound lipid is shown in Fig. 3.17.

Bibliography

For the structure of biological macromolecules, consult a standard text-book of biochemistry.

Barman, T. E. (1969), The Enzyme Handbook – Vols. 1 and 2, Springer, Vienna. *Summarized information on all known enzymes.*

Methods in Enzymology.
Recent volumes cover in more detail most of the hydrolases of interest.
Barret, A. J. (1977) (Edit.) *Proteinases in Mammalian Cells and Tissues*, North
    Holland, Amsterdam.
An excellent comprehensive treatise.
Barrett, A. J. and Heath, M. F. (1977), in *Lysosomes, a laboratory handbook*, 2nd
edition. (Ed.) J. T. Dingle, pp. 19–145, North Holland, Amsterdam.

4 Lysosomes; a specialized degradative organelle in eukaryotic cells

4.1 Characteristics of lysosomes

In this chapter we make a first approach to the question of the cellular
machinery involved in the chemical breakdown steps we have described
already. Although degradative events involving macromolecules as substrates occur throughout cells, eukaryotic cells possess a specialised
degradative organelle, the lysosome. Here we survey lysosomal function
to provide a background for subsequent discussion of integration and
control of breakdown. Plates 1 and 2 show lysosomes in cells, and purified lysosomes.

Lysosomes are intracellular vesicles, often about 0.5μ in diameter,
surrounded by a membrane, and containing hydrolytic enzymes, which
are most active under acid conditions. Although they are morphologically
extremely heterogeneous, and undergo diverse tranformations by fusing
with other cellular membranes (see Section 4.2), they share several
functional characteristics.

Firstly, the membrane confers on the enclosed enzymes the property
of 'latency'. When a homogenate of a tissue containing lysosomes is
prepared, and acid hydrolases are assayed under osmotically protected
conditions (usually osmotic protection is provided by the presence of
0.25M-sucrose, which is isosmotic with biological structures and does not
easily pass through membranes), a low activity is found. But, if the
suspension of tissue homogenate is frozen and thawed several times, or
subjected to mechanical stress (with a Waring blendor or a sonicator), or
treated with a detergent (such as Triton X-100), much more activity is
revealed. This activity is said to have been 'latent' in the original
homogenate. It was this property which led to de Duve 1955 to the discovery
of lysosomes.

Latency is due to the presence of an intact membrane around the
lysosomal enzymes. The membrane limits the access of low molecular
weight compounds (such as substrates) to the interior of the lysosomes,

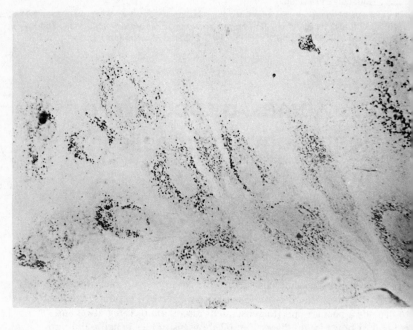

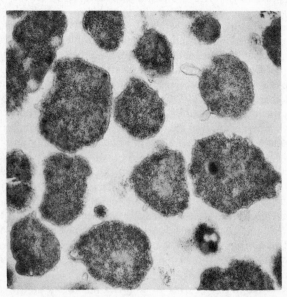

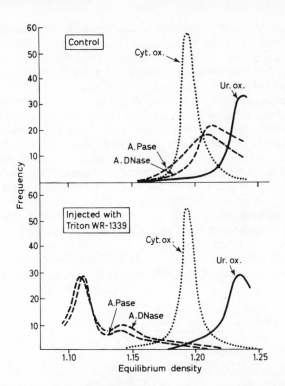

Fig. 4.1 Distribution of organelles of rat liver homogenates centrifuged on aqueous sucrose density gradients. The organelles are detected by marker enzymes. The 'injected' preparation was from a rat injected four days previously with Triton WR-1339. The 'control' rat was not specially treated. Triton brings about a selective shift of lysosomes (acid phosphatase, acid DNase) to lower density positions of the gradient. Mitochondria (cytochrome oxidase) and peroxisomes (urate oxidase) are not affected by the detergent. Note the slight differences in sedimentation behavior of the two lysosomal hydrolases; this may reflect some form of heterogeneity in the lysosome population. (From De Duve, C., 1965: *Harvey Lects.* **59**, 49–87.)

Plate 1. Localization of cathepsin D in rabbit fibroblasts, revealed by means of a specific antiserum to cathepsin D. Particulate staining indicates the lysosomal sites of activity. Magnification x1000 approx. (By kind permission of Dr. A. R. Poole.)

Plate 2. Purified iron-loaded rat liver lysosomes. The lysosomal membrane is clearly visible, and shows some irregularities of contour which may be corollaries of the lysosomal capacity for fusion and invagination, and possibly, for budding outwards. The granular material within the lysosomes is the stored iron-containing materials (primarily ferritin). Magnification x100 000. (Picture kindly supplied by Dr Hans Glaumann)

33

and completely prevents the escape of the large molecules, such as the enzymes, from lysosomes. Hence, when the membrane is intact, little enzyme—substrate interaction can take place and so little enzyme activity is detected. When the membrane is disrupted, interaction is no longer limited, and full activity is detected.

The membrane is also responsible for the fact that lysosomes can be sedimented by centrifugation. Indeed, de Duve first showed that acid hydrolases were associated with a particle other than mitochondria by showing that the sedimentability properties of the enzymes were different from those of mitochondrial enzymes. Of course, the morphological and functional heterogeneity mentioned already, has concomitant heterogeneity in sedimentability, and this creates problems in the isolation of lysosomes. A further difficulty in this respect is that lysosomal sedimentability is very similar to that of mitochondria, and the particles overlap each other extensively. In addition, there are far more mitochondria than lysosomes.

These difficulties can be overcome by loading lysosomes *in vivo* with particles or compounds of extremely high or low density (high: colloidal

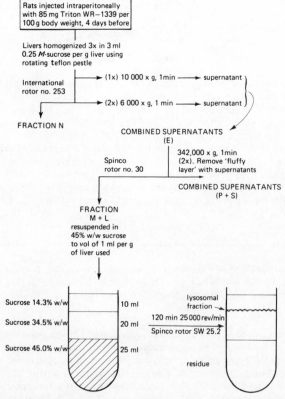

Fig. 4.2 The isolation of Triton-WR-1339 loaded rat liver lysosomes.

Table 4.1 A selection of lysosomal enzymes.

Name (synonyms)	Enzyme Commission number	Comments	pH optimun
Oxidoreductases			
NADPH$_2$ oxidase	1.6.2.–?		
Peroxidase	1.11.1.7		5.0–6.0
Hydrolases acting on carboxylic esters			
Phospholipase A$_2$ (phospho-lipid 2-deacylase)			4.5
Cholesterol esterase	3.1.1.4		4.0–4.5
Phospholipase A$_1$ (phospholipid 1-deacylase)	3.1.1.13		
	3.1.1.32		3.5–4.0
Hydrolases acting on phosphoric monoesters			
Acid phosphatase	3.1.3.2		3.0–6.0
Hydrolases acting on phosphoric diesters			
Deoxyribonuclease II	3.1.4.6	Endo-cleavage of double-stranded DNA, often both strands at the same point, leaving 3′-phosphate termini.	3.8–5.5
Ribonuclease II	3.1.4.23	Endo-cleavage of RNA, leaving 3′-phosphate termini, 2′ 3′-cyclic phosphates being intermediates.	5.4–6.7
Hydrolases acting on sulphuric esters			
Sulphatase A (arylsulphatase A)	3.1.6.1		
Cerebroside sulphatase	3.1.6.8	Some or all forms act on cerebroside 3-sulphate and ascorbic acid 2-sulphate	4.5–4.8

Table 4.1 (contd.)

Name (synonyms)	Enzyme Commission number	Comments	pH optimum
Hydrolases acting on glycosides			
Lysozyme (muramidase)	3.2.1.17	Cleaves the (1β-4) linkage of N-acetyl-6.2 muramic acid to N-acetylglucosamine in the polysaccharide component of the cell walls of some bacteria, and the N-acetylglucosaminyl-(1β-4)-N-acetylglucosamine linkage in chitin.	6.2
Neuraminidase	3.2.1.18	Cleaves non-reducing terminal α-glycosidic linkages of N-acetyl-neuraminic acid in glycoproteins and glycolipids.	4.0–4.5
β-N-Acetylglucosaminidase (β-N-acetylhexosaminidase)	3.2.1.30 3.2.1.52		3.7–5.0
β-Glucuronidase	3.2.1.31		4.3–5.0
Hyaluronate endogluco-saminidase (hyaluroni-dase, hyaluronoglucos-aminidase)	3.2.1.35		3.5–4.1
Hydrolases cleaving peptide bonds near the ends of polypeptides: exopeptidases			
Lysosomal aminopeptidase (amino acid naphthyla-midase, cathepsin H)	3.4.11		Various
Hydrolases cleaving peptide bonds away from the ends of polypeptides: endopeptidases			
Acrosin	3.4.21.10	Present in sperm	8.0
Lysosomal elastase (leuco-cyte elastase)	3.4.21.11	Found in only a few types of cells such as the neutrophil leukocyte serine proteinases	7.5

Enzyme	EC number	Description	pH
Cathepsin G (chymotrypsin-like enzyme)	3.4.21.20		
Cathepsin B (cathepsin B', cathepsin B1)	3.4.22.1	A major thiol proteinase	3.5–6.0
Cathepsin L	3.4.22	A thiol proteinase	5.0
Cathepsin D	3.4.23.5	A major carboxyl proteinase	2.8–5.0
Cathepsin E	3.4.23	A carboxyl proteinase	2.5
Granulocyte collagenase	3.4.24.7	A metallo proteinase	7.5
Hydrolases acting on amide bonds other than peptides			
Aspartylglucosylaminase	3.5.1.26	Crucial for glycoprotein degradation (q.v.).	5.1–7.6
Hydrolases acting on acid anhydrides			
Nucleoside triphosphatase (acid pyrophosphatase)	3.6.1.15	Cleaves ATP and other nucleoside triphosphates to yield the disphosphate.	4.0–5.2
Hydrolases acting on nitrogen–sulphur bonds			
Heparin sulphamatase (heparin sulphamidase or *N*-sulphatase)	3.10.1	Liberates sulphate ion from *N*-sulphoglucosaminyl residues in heparin and heparan sulphate.	4.5–5.1

Modified from Barrett and Heath, 1977 (see Bibliography)

Table 4.2 Lysosomal storage diseases (Modified from Neufeld et al., 1975: see Bibliography.)

Disorder	Enzyme deficiency	Metabolites primarily affected
Mucopolysaccharidoses		
Hurler and Scheie syndromes	α-L-Iduronidase	Proteoglycan components
Hunter syndrome	Iduronate sulphatase	
Sanfilippo syndrome		
A subtype	Heparan N-sulphatase	
B subtype	N-Acetyl-α-glucosaminidase	
Maroteaux-Lamy syndrome	N-Acetylgalactosamine sulphatase (arylsulphatase B)	
β-Glucuronidase deficiency	β-glucuronidase	
Morquio syndrome	Uncertain	
Sphingolipidoses		
GM₁ gangliosidosis	β-Galactosidase	GM₁ ganglioside, fragments from glycoproteins and certain complex lipids
Krabbe's disease	β-Galactosidase	
Lactosylceramidosis	β-Galactosidase	
Tay-Sachs disease	Hexosaminidase A	
Sandhoff's disease	Hexosaminidases A and B	
Gaucher's disease	β-Glucosidase	
Fabry's disease	α-Galactosidase	
Metachromatic leukodystrophy	Arylsulphatase A	
Niemann-Pick disease	Spingomyelinase	
Farber's disease	Ceramidase	

Disorders of glycoprotein metabolism		
Fucosidosis	α-L-Fucosidase	Fragments from glycoproteins
Mannosidosis	α-Mannosidase	
Aspartylglycosaminuria	Amidase	
Other disorders with single enzyme defect		
Pompe's disease	α-Glucosidase	Glycogen
Wolman's disease	Acid lipase	Cholesterol esters, triglyceride
Acid phosphatase deficiency	Acid phosphatase	Phosphate esters
Multiple enzyme deficiencies		
Multiple sulphatase deficiency	Sulphatases (arylsulphatase A, B, C: steroid sulphatases; iduronate sulphatase; heparan N-sulphatase)	Steroid sulphatases; proteo proteoglycans
I cell disease and pseudo-Hurler polydystrophy	Almost all lysosomal enzymes except protein-ases deficient in cultured fibroblasts; present extracellularly	Proteoglycans and complex lipids
Disorders of unknown origin		
Cystinosis	Accumulation of cystine in lysosomes	Cystine
Mucolipidoses I, IV	Ultrastructural evidence of lysosomal storage	Unknown

gold; low: the non-lytic detergent, Triton WR-1339) so that the density of lysosomes is shifted away from that of mitochondria and other potential contaminants. Fig. 4.1 shows the effect on lysosomal density of the loading *in vivo* with Triton WR-1339, and the lack of effect on other organelles. An alternative, is to obtain a mixed preparation of mito-chondria and lysosomes, and to alter the mitochondrial density by incu-bation with a substrate for a mitochondrial enzymes, and converting the product of the reaction into a dense precipitate. Thus selective alteration of mitochondrial density can be achieved, and so the lysosomes purified. An example of one of the standard isolation procedures for rat liver lyso-somes is given in Fig. 4.2, which illustrates the considerable effect involved!

As has already been mentioned, lysosomes contain a most impressive array of hydrolases, and seem to be capable of degrading all known cellular substrates. An outline classification of lysosomal enzymes is given in Table 4.1, with only a selection of examples. The demonstration, by enzyme staining techniques, of the presence of such enzymes in a vesicle is a good means of identifying the vesicle as a lysosome. Plate 1 shows a demonstration of cathepsin D (a major lysosomal protein) by means of a specific antibody to the enzyme. This antibody reacts with no other protein, and so can be used to demonstrate the presence and localization of the enzyme, if the antibody is 'labelled' in some detectable way. In this case the antibody was labelled with a fluorescent dye.

The potency and importance of lysosomal enzymes in the degradation of many substrates is revealed by the profound effects of the genetic deficiency of single lysosomal enzymes in the storage diseases, which are usually lethal in early life. As a result of the enzyme deficiency lyso-somes are no longer capable of digesting certain of their normal sub-strates, which therefore accumulate in lysosomes giving a character-istic morphology of distended organelles, often containing whorls of lipids. Table 4.2 shows some of the characterized storage diseases.

The importance of the membrane in sequestering the hydrolases, and thus avoiding rampant degradation by them, is obvious. But the mem-brane also assists in maintaining an intralysosomal environment optimal for lysosomal enzyme activity, characterised by low pH. This is probably mainly through the maintenance of a Donnan equilibrium across the membrane. Because there are substantial amounts of large molecular weight negatively charged compounds (such as glycoproteins with negatively charged neuraminic acid residues, and negatively charged phospholipids) retained in the lysosome by the membrane, there is an associated accumulation of univalent cations, primarily protons, inside the lysosome. This mechanism seems to result in a passive pH gradient of about 1.5 pH units across the membrane.

This can be measured by studying the distribution of weak bases *in vitro* in lysosomal suspensions. Such compounds are accumulated by lysosomes because of the availability of internal protons (see Fig. 4.3). Uncharged small molecules can penetrate biological membranes, though at varying speeds, but charged forms cannot. Thus intralysosomal

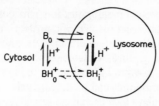

Fig. 4.3 The intralysosomal pH and the accumulation of weak bases. The permeating weak base (B) is in equilibrium with its protonated form (BH^+), which permeates very poorly (-----). If the H^+ concentration within the lysosomes is higher than that in the cytosol, the BH^+ form will reach much higher concentrations within than without. And thus the intralysosomal total base concentration ($[B_1] + [BH_1]$) will be greater than the cytosol ($[B_0] + [BH_0]$), as $B_0 = B_1$: the ratio $[B_1 + BH_1]/[B_0] + [BH_0]$ will increase as the intralysosomal $[H^+]$ rises, and can be used to determine the latter.

protonated molecules of a weak base are trapped by virtue of their positive charge; and the quantity of base accumulated can be used to measure the pH gradient between the inside and outside of the lysosomes. Some compounds are accumulated even more than the pH gradient itself would allow. In some cases such as the fluorescent dye, acridine orange, this is a result of direct binding to membrane components. But there remain certain cases of 'excess' accumulation of, compounds, and it is possible that lysosomes may have a proton pump, using energy from hydrolysis of ATP, which increases lysosomal acidity still further. The intralysosomal hydrolyses themselves generate protons, so it is quite possible that pH gradients of more than 1.5 units are maintained *in vivo*.

The restricted permeability of membranes for macromolecules, not only contains the lysosomal enzymes, but also macromolecular substrates undergoing digestion in lysosomes. Only the low molecular weight products of digestion (such as free amino acids, sugars and nucleotides) are able to penetrate the membrane, and thus to contribute to cellular nutrition. So lysosomal degradation is normally an all or none process: once within a lysosome, a substrate can only be completely degraded. Minor exceptions occur during cell death, with concomitant breakage of lysosomal membranes, and a significant exception is probably involved in the secretion of certain proteins from cells. As noted earlier, such proteins are often synthesized in 'pro-forms', which undergo limited proteolysis to produce the functional molecule. It seems that the secretory vesicle may be a modified lysosome (see Section 4.2.) and yet the secreted molecules only undergo limited cleavage. This may be a consequence of the kinetics and specificity of proteolysis: the time from acquisition of lysosomal proteinases to secretion (and thus cessation of degradation) is short, and the pro-peptide may be the best substrate available for the enzymes present (which may be only a small selection of lysosomal proteinases).

Permeability of lysosomal membranes to a particular compound can be studied *in vitro* by assessing the osmotic protection provided by that compound: non-penetrating compounds protect, preventing water influx and concomitant rupture; while penetrating ones do not, since they enter and so cause a water influx leading to lysis. Non-penetrating and non-degradable compounds can be detected also using living cells: they are able to accumulate in cells and cause a distension of the lysosomal system which is easy to observe by phase contrast microscopy. This has been shown with several compounds, such as sucrose and other non-digestible disaccharides. The accumulation of these disaccharides, which are too large to penetrate membranes, depends on their entry by endocytosis (unlike the permeating weak bases).

Thus the lysosomal membrane performs the balanced roles of constraining the lysosomal enzymes while allowing transit of some low molecular weight molecules, such as the products of digestion. It probably also performs more subtle regulatory activities: these may include controlled transport of ions, and controlled entry of macromolecules which bind to the exterior of the lysosomal membrane and are then internalized by membrane invagination (see Chapter 6).

4.2 The dynamics of the lysosomal system

The synthesis of lysosomal components has so far received little study. But it is clear that lysosomes are incapable of protein synthesis, and that lysosomal proteins are synthesized on cytoplasmic ribosomes. The roles of the free and membrane-bound ribosomes have not yet been elucidated, but it is clear that one translocation route for lysosomal enzymes is from rough to smooth endoplasmic reticulum, to Golgi apparatus, to lysosomes. The packaging of enzymes following this route occurs in the Golgi, but it is not clear whether cisternal elements or only peripheral elements are involved. An alternative route from endoplasmic reticulum to lysosomes, bypassing the Golgi, has been proposed by Novikoff. It is embodied in his GERL(Golgi-associated endoplasmic reticulum giving rise to lysosomes) concept, in which specialized regions of smooth endoplasmic reticulum, adjacent to, but distinct from, Golgi, give rise to lysosomal vesicles. The GERL hypothesis is supported by elegant and extensive enzyme cytochemistry by Novikoff and his associates. There are no kinetic biochemical data which yet confirm the GERL route, or allow an assessment of the relative importance of the two routes. The GERL system is shown schematically in Fig. 4.4.

The membrane of the lysosomal system is maintained and exchanged by several different mechanisms (illustrated in Fig. 4.5). First of all, there is production of lysosomal vesicles in the Golgi/GERL region. These lysosomes are termed Primary lysosomes (see Fig. 4.5. for lysosomal terminology) as they have so far received no substrate from outside the lysosome. Primary lysosomes fuse with many other vesicles, and, on occasion, with the plasma membrane (resulting in secretion of lysosomal enzymes). The products of the former fusions are termed *secondary* lysosomes since they do contain substrate from outside the lysosomes.

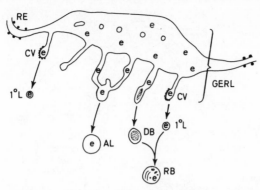

Fig. 4.4 The GERL system (after Novikoff, in Hers and van Hoof, (1973): see Bibliography). The GERL is a specialized region of smooth endoplasmic reticulum, and probably gives rise to three types of lysosome. (1) The coated vesicles (CV) which probably form primary lysosomes ($1°L$): (2) via a type of autophagy it produces autolysosomes (AL): (3) dense bodies (DB), which may contain lysosomal enzymes initially and which can also fuse with other lysosomes, eventually giving rise to residual bodies (RB) containing some remnants of the digestive processes.

Secondary lysosomes are just as capable of fusing with other vesicles as are primary ones, so that every lysosome is in continuity with every other by means of such fusion reactions, and new particles arriving within the lysosomal system are rapidly spread throughout it. Lysosomes do not normally fuse with endoplasmic reticulum which is intact, with mitochondria or with nuclei. They may however interact with all these structures during autophagy (see later). The vesicles which fuse with lysosomes result from two main processes, endocytosis (which is relevant to the degradation of extracellular materials) and autophagy (which is concerned with the degradation of intracellular structures).

Endocytosis is a process by which extracellular molecules enter a cell within vesicles which form at the plasma membrane. It comprises two main categories: phagocytosis and pinocytosis. In the former, particulate material such as foreign organisms (bacteria) are endocytosed by means of a large vacuole (the phagosome). In contrast, pinocytosis is the endocytosis of small objects, usually soluble molecules, though sometimes colloids, and of material which binds to the plasma membrane. The vesicles formed are much smaller than phagosomes and are termed pinosomes. Both phagosomes and pinosomes are intially pre-lysosomes, since they lack lysosomal enzymes, but soon fuse with lysosomes to become secondary lysosomes. In fact, such fusion may even begin before phagocytic vacuoles have completely closed, so that some secretion of lysosomal enzymes is often concomitant with phagocytosis.

Phagocytosis requires the participation of microfilaments, contractile filaments present in most cells, and can be prevented by the fungal metabolite, cytochalasin B, which seems to paralyse microfilaments.

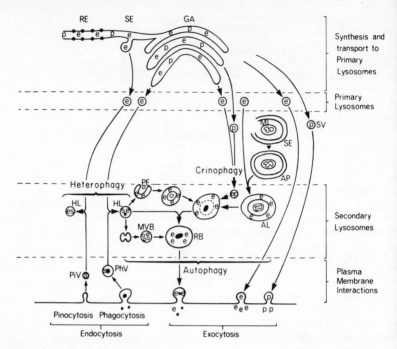

Fig. 4.5 Lysosomal fusions. The general categories primary lysosomes (containing lysosomal enzymes, but not having received substrates), and secondary lysosomes (in which substrates are undergoing enzymic attack) are distinguished. In addition, the term prelysosome covers PiV, PhV and AP: vesicles containing substrates but awaiting the entry of lysosomal enzymes. Several alternatives for the terms used here are extant (see de Duve and Wattiaux, 1966) and the most important are: autophagic vacuole or cytosegresome, which cover AP and AL together; and endocytic vesicle and heterophagosome which cover PiV and PhV together. It should be emphasized that not all the lysosomal transformations indicated here necessarily occur in every cell type and that some lysosomal activities are omitted (e.g. penetration of the nuclear membrane). But in general there is a remarkably continuous exchange between all members of the lysosomal system: thus residual bodies can still fuse, with most other types of lysosome, and there is free fusion between vesicles produced by autophagy and those from heterophagy. *Within membranes:* e-lysosomal enzymes; p-secretory proteins; s-soluble substrates for lysosomal digestion; particulate substrates are represented as solid areas. *Outside membranes:* RE-rough endoplasmic reticulum; SE-smooth endoplasmic reticulum; GA-Golgi apparatus; SV-secretory vesicle; AP-autophagosome; AL-autolysosome; PE-peroxisome; MVB-multivesicular body; RB-residual body; PiV-pinocytic vesicle; PhV-phagocytic vacuole; HL-heterolysosome; M-mitochondrion.

44

Microtubules may be important in phagocytosis by some cells, but more probably, they are involved in control of movement of secretory vesicles. Pinocytosis takes two distinct forms: macropinocytosis utilises microfilaments and gives rise to vesicle visible at the light microscope level, while micropinocytosis produces smaller vesicles and seems to be independent of microfilaments. The energy requirements of the two processes differ also: micropinocytosis shows only slight requirements, while macropinocytosis is clearly dependent.

The remaining major source of lysosomal membrane is autophagy, which takes several forms. In classical autophagy, recognisable remnants of intracellular organelles are seen surrounded by one or two membranes. These surrounding membranes usually originate from endoplasmic reticulum, and the resultant vacuole fuses with lysosomes, and the inner membrane together with the contained organelle (such as a mitochondrion) undergoes digestion. But lysosomal autophagy occurs also: a lysosome invaginates to form a lysosome containing an inner membrane (or several internal small vesicles) surrounded by the outer membrane. The internal membrane surrounds cytosol material which comes into contact with lysosomal enzymes when the inner membrane is broken during digestion. Thus lysosomal membrane and cytosol material is digested by this mechanism, which will be discussed further later (Chapter 6). More subtle forms of autophagy have been recognized relatively recently by Novikoff ('Microautophagy'), which involve endoplasmic reticulum regions acting on a very small scale.

Several cell types are capable of secreting lysosomal enzymes (without concomitant cell death), and this of course, results in the lysosomal system losing membrane to the plasma membrane. Prominent amonst these are the phagocytic cells such as blood leukocytes. These are attracted by chemotactic factors to the sites of infection and immunological activity. Neutrophilic leukocytes respond to the presentation of foreign particles, and of immunological stimuli (such as immune complexes and activated complement components [see Chapter 5]) by a very rapid secretion of a large part of their lysosomal enzymes. This is usually a once-only occurrence, as the life span of the neutrophilic is short. In contrast, the blood monocyte, which when attracted to a site of infection matures to the macrophage, can secrete lysosomal enzymes over very long periods. It is thus probably an important agent in chronic (long-lasting) inflammation, where excessive tissue destruction continues for long periods.

The prolonged secretion by macrophages indicates that secretion is not necessarily simultaneous with phagocytosis. Indeed, several mechanisms of secretion are suspected (Fig. 4.6): two are concomitant with phagocytosis, one a consequence of intracellular storage of phagocytosed materials, and one induced by surface ligands such as immunological components or hormones. Secretion in response to hormonal stimuli is seen in resorbing bone, while cartilage cultures secrete in response to Vitamin A or to the storage of sucrose in their lysosomes. Thus secretion is by no means restricted to the professional phagocytes.

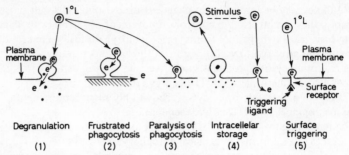

Degranulation (1)   Frustrated phagocytosis (2)   Paralysis of phagocytosis (3)   Intracellelar storage (4)   Surface triggering (5)

Fig. 4.6 Exocytosis of lysosomal and other enzymes accompanying phagocytic processes and other surface stimulations. In mechanisms (1, 3, 4), normal particulate stimuli are applied. In (1), degranulation of primary lysosomes into the forming phagosome occurs to some extent before it has closed. In (2) and (3), phagocytosis is initiated, but not completed: in (2) this is because a phagocytosable material has been insolubilized on a large solid support too big to enter the cell, and in (3) because phagocytosis has been inhibited, for instance by cytochalasin B. In (4), intracellular storage of non-digestible material somehow stimulates the release of neutral proteinases from probably non-lysosomal granules. In (5) a ligand binds to a membrane receptor, and induces release (endocytosis of the ligand may or may not occur at the same time time). $1°L$ – primary lysosome; e – enzyme; NPG – neutral proteinase containing granule.

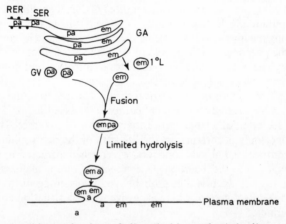

Fig. 4.7 A possible mechanism of albumin biosynthesis by liver proalbumin (pa) is formed on rough endoplasmic reticulum (RER) and transferred via smooth endoplasmic reticulum (SER) and the Golgi apparatus (GA) to Golgi vesicles (GV). These fuse with primary lysosomes ($1°L$) containing some membrane-bound proteinases (em) which cleave the proalbumin molecule to release a terminal peptide and native albumin. The peptide seems to be degraded and albumin is released by exocytosis, while the proteinases may remain bound to the plasma membrane, and subsequently be re-endocytosed. Abbreviations as in Fig. 4.5 except: *Within membranes:* em – membrane-Bound lysosomal enzyme.

There are other secretory vesicles besides lysosomes of course. Secretion of serum proteins by liver has already been outlined. It seems that the supply of proteinases for cleavage of the pro-forms of such protein may be made by a fusion reaction between a lysosome (possibly of a rather specialized kind) and the secretory vesicle, just like the other fusions lysosomes undergo (Fig. 4.7). But there is little evidence that the secretion of serum proteins, or of hormones showing pro-forms, is accompanied by lysosomal enzyme release: this may be because the enzymes are retained on the cell surface, or because only a small amount of enzyme is released and it has not yet been detected.

It is also notable that amongst other non-lysosomal secretory vesicles there are some which contain degradative enzymes. For instance, lysozyme is secreted continuously by macrophages, in contrast to lysosomal enzymes which are only released when the cells are suitably stimulated. Macrophage lysosozyme is probably vesicle-bound, and clearly the vesicle must be distinct from lysosomes, or at least must be an extremely specialized lysosome. In contrast, neutrophil lysosomes possess lysozyme. Macrophages also have other categories of secreted products: several neutral proteinases (such as a plasminogen activator, and a 'specific collagenase [which produces only the limited cleavage: see Table 4.2]) are released on appropriate stimulation. Some of the stimuli are not effective in inducing lysosomal release (e.g. latex particles) while others are. So these enzymes seem also to be non-lysosomal. Other cells, such as fibroblasts, also secrete neutral proteinases in response to stimuli; and of

Table 4.3 Some secretory products of macrophages

Product	Stimulus for production or release
Lysosomal hydrolases	Zymosan
	Dental plaque
	Group A streptococcal cell walls
	Immune complexes
	Asbestos
	Activated complement
	Human lymphocyte activation products
	Carrageenan
Plasminogen activator	Thioglycollate *in vivo*
	Phagocytosis *in vitro*
Collagenase	Endotoxin
Lysozyme	None required
Pyrogen	Endotoxin
	Phagocytosis
Prostaglandins	Lymphocyte activation products
Complement components	Requirements not clear
α_2-macroglobulin (a general proteinase in inhibitor)	Requirements not clear

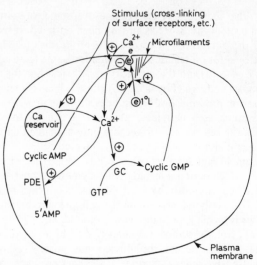

Fig. 4.8 A simplified scheme of the mechanism of induction of exocytosis. Calcium influx or release from an intracellular reservoir is consequent upon interaction of the stimulant with the plasma membrane. The resultant rise in intracellular calcium concentration initiates exocytosis and facilitates production of cyclic GMP by Guanyl cyclase (GC), and removal of cyclic AMP by phosphodiesterases (PDE). Cyclic GMP stimulates the exocytosis, while the inhibitory effect of cyclic AMP is minimized by its breakdown. Mechanisms for recovery of calcium and cyclic nucleotide levels to normal are not indicated. Other abbreviations: $\oplus$ and $\ominus$, stimulation and inhibition respectively, by a metabolite; $1°L$ – primary lysosome; e – lysosomal enzymes.

course many cells synthesise and secrete other proteins. The diversity of secretions of which the macrophage is capable is shown in Table 4.3.

The mechanism of induction of triggered secretions (as opposed to that of continuous secretions, such lysozyme from macrophages, and serum proteins from liver) is complex, but calcium fluxes are crucial. Some of the basic interactions are shown in Fig. 4.8. The involvement of degradative steps in triggered responses is discussed later (Chapter 5).

The neutral proteinase containing granules seem so far to be quite analogous to lysosomes, merely lacking the acid hydrolases; but unlike lysosomes they seem normally to be produced only when their secretion is required. The combined extracellular release of both categories of enzyme would make available a very wide range of extracellular enzymes. The relative importance of the different categories of enzyme in extra-cellular degradation is discussed in Chapter 5, together with their possible roles in initiating the chain proteolytic cascades of blood.

Thus lysosomes gain membrane from autophagy and endocytosis, and lose it by fusion with plasma membrane during secretion, and by genuine breakdown. The characteristics of turnover of lysosomal components are described later (Chapter 5).

4.3 Degradative roles of lysosomes

Here we discuss the involvement of lysosomes in various degradative processes: the rates and control of such processes are covered later (Chapter 5 and 6).

It is clearly established that endocytosed materials (bacteria, particles, etc) join the lysosomal system, and undergo digestion within it. But there are certain intracellular parasites (such as *Toxoplasma gondii*) which avoid fusion by mechanisms which are incompletely understood, but which require the organism to be alive. Of course, some materials are indigestible. In addition, there are circumstances in which bacteria (which are normally mostly digestible) are retained in cells: these may reflect cellular defects in the host cells (as is possibly the case in phagocytes in Crohn's disease, which involves considerable tissue damage in the intestine), and may result in secretion of lysosomal enzymes and subsequent tissue damage. But in general, endocytosed materials are completely degraded, and their chemical constituents then made available to cellular nutrition. In some cases, the material requires pretreatment extracellularly before it can be endocytosed. For instance, connective tissue comprises polymeric molecules associated in fibrous aggregates much bigger than cells. Thus extracellular proteinases such as the specific collagenases first cleave the intact fibrils to release smaller macromolecules which can dissociate from the aggregates and be endocytosed.

The evidence that lysosomes are responsible for the digestion of endocytosed materials is very strong. There is histochemical evidence that substrate and lysosomal enzymes are associated with the same vesicles, there is biochemical evidence that the substrates initially sediment with lysosomes before undergoing degradation, and it was shown most elegantly by Barrett and Dingle that when inhibitory antibodies specific for cathepsin D (one of the major lysosomal proteinases) were endocytosed at the same time as a substrate, haemoglobin, the subsequent intracellular degradation of haemoglobin was retarded. Companion experiments showed that haemoglobin and cathepsin D were localized in the same particles.

The attempts to prove that extracellular lysosomal enzymes have roles in connective tissue catabolism have been more equivocal. A long series

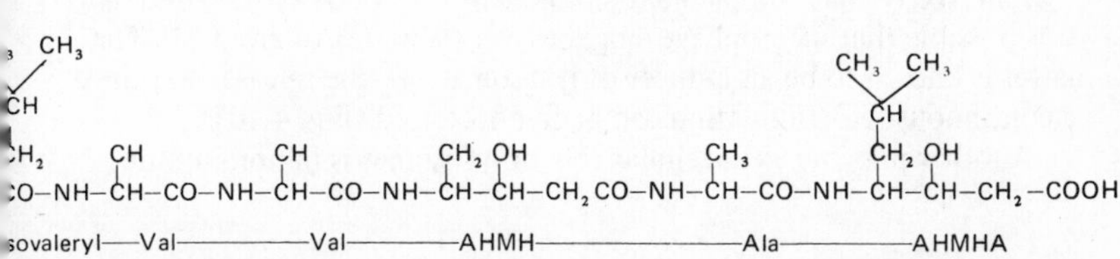

Fig. 4.9 Pepstatin, the potent peptide inhibitor of cathepsin D. This is isovaleryl-L-valyl-L-valyl-4-amino-3-hydroxy-6-methylheptanoyl-L-alanyl-4-amino-3-hydroxy-6-methylheptanoic acid. The AHMH residue is not a normal residue in proteins and is probably crucial for the inhibiting effect.

49

of studies showed that cathepsin D is released from cells during cartilage degradation, and that autolysis of dead cartilage could be inhibited by the specific inhibitor of cathepsin D, pepstatin (Fig. 4.9) obtained from *Actinomyces* cultures. But no direct evidence that cathepsin D is important in the breakdown of living cartilage has been obtained. Since the enzyme has an acid pH optimum, and yet physiological extracellular pH is near neutrality, it is not obvious that cathepsin D could function, unless in a local microenvironment, with pH very different from bulk pH. It is known that charged solids induce a local environment very different from the bulk medium, as a result of electrostatic and other interactions, so this is feasible. However, the role of the acid hydrolases remains to be proved, although the correlation between exaggerated tissue damage in inflammation, and the release of lysosomal enzymes is suggestive.

Perhaps the best candidates for participation in extracellular digestion are the neutrophil lysosomal serine proteases, cathepsin G and elastase, and the other neutral proteinases from macrophages and fibroblasts. Since their pH optima are near neutrality the necessity for postulating action within a microenvironment is avoided.

As has been mentioned already, limited proteolyses are involved in the activation of several physiological cascades reactions in extracellular fluids. In most of these it is known that lysosomal enzymes are capable of initiation under controlled experimental conditions. But again, the conditions required make it unclear whether these reactions have physiological relevance. One possible exception is in the production of kinins. Greenbaum has presented evidence using pepstatin, that activation by cathepsin D does have a significant role *in vivo*. (Fig. 4.10). The only obvious alternative interpretations are that the effect of the inhibitor was an indirect one, or that some carboxyl proteinase other than cathepsin D, with a neutral pH optimum, was involved.

Again, the neutral proteinases from lysosomal and other granules may provide alternative candidates for the activation of such systems. It is particularly interesting that several of the activated components of the complement cascade are inducers of lysosomal enzyme release from macrophages and neutrophils: a continuous re-stimulation can thus be envisaged. In the case of a macrophage in a chronic inflammatory lesion, it seems likely that complement component C3 can be synthesized, and it is possible that macrophage enzymes can cleave C3 to give C3b. The latter is known to be an inducer of lysosomal enzyme release, and thus a continuous self-stimulation can be hypothesized (Fig. 4.10).

Another possible extracellular role of lysosomes is in non-immuno-

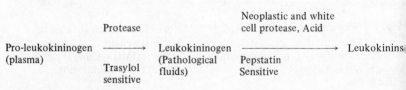

Fig. 4.10 Leukokinin forming system.

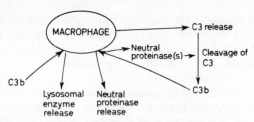

Fig. 4.11 The macrophage probably secretes C3 continuously in an inflammatory lesion; when C3b is present, secretion of enzymes is induced and amongst the enzymes may be one which cleaves C3 to produce more C3b and thus sustain the enzyme release. It is interesting that C3 seems to be synthesized as a proform (like pro-insulin), and thus it may undergo both intracellular and entracellular clearage.

logic killing of tumour cells; this may involve direct secretion of lyso-somal enzymes into target cells. It is inhibited by membrane stabilisers, by pretreatment of the macrophages with Trypan Blue (which inhibits several lysosomal enzymes), and by simultaneous presentation of phago-cytosable particles. All these observations are suggestive of lysosomal participation. Lysosomal enzymes are not released into the medium during killing, so one has to postulate direct release into the target cell cytoplasm. A similar mechanism may be involved in other forms of killing (by lymphocytes for instance).

Just as it has been very difficult firmly to implicate lysosomal enzymes in extracellular degradation, so has it been difficult to demonstrate their role in the degradation of endogenous intracellular materials (as opposed to endocytosed molecules). On the other hand, the storage diseases give strong indication of the range of substrates normally degraded by lyso-somes. Characteristics of several storage diseases have been summarized earlier (Table 4.2). Some of the stored materials may originate extra-cellularly (from normal components or from dying cells, etc.) but most probably originate to some degree from within the cells.

There are as yet no known storage diseases in which proteinases are deficient, and proteins stored in lysosomes. Yet evidence is now strong that lysosomes contribute to the degradation of intracellular proteins. The lack of proteinase storage diseases may be because they are highly lethal, and cause foetal abortion (whereas other storage diseases only lead to death in childhood).

The initial evidence for turnover of intracellular proteins in lysosomes, was an extensive set of data concerning the state of the liver lysosomes in perfused liver (in respect of size, fragility, and number) in relation to the ongoing rate of proteolysis. This rate is sensitive to insulin and amino acids (inhibitory) and to glucagon (stimulatory). Perfusion itself is a stimulus to proteolysis. Increased degradation is accompanied by enlarge-ment of lysosomes, increased volume of the lysosomal system, and exaggerated lysosomal lability. Parallel observations have been made on perfused hearts.

There was also evidence from the effects of weak bases which accumulate in lysosomes as described earlier: these agents inhibit protein breakdown, partly as a result of raising the intralysosomal pH (by binding protons), and in the case of chloroquine, perhaps also by inhibiting cathepsin B. Of course these agents will be present in all parts of the cells, and may have multiple actions. So their effects on proteolysis cannot be easily interpreted.

This indirect evidence of lysosomal involvement in proteolysis, was later confirmed by several kinds of experiment using the only diagnostic inhibitor of a lysosomal proteinase, pepstatin. This is diagnostic because it is a specific inhibitor of carboxyl proteinases, and because in many cell types the only carboxyl proteinase seems to be lysosomal cathepsin D. The first set of experiments used the perfused liver system, which had been extensively characterized, as indicated already. Pepstatin does not seem to penetrate cellular membranes directly, so it was found that free pepstatin does not inhibit proteolysis over a period of two hours. Later experiments with isolated cells in culture show that pepstatin does progressively inhibit, over a matter of about 24 hours, probably reflecting slow endocytosis which then allows access to the lysosomes. But the lack of permeability of pepstatin through cellular membranes was exploited by entrapping it within liposomes (Fig. 4.12) which

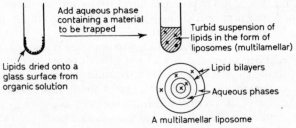

Fig. 4.12 Formation of liposomes, continuing a non-penetrating solute (X). On addition of the aqueous phase, the liposomes form spontaneously, and entrap some aqueous fluid containing X. Molecules of X which remain outside the liposomes in the bulk phase may be removed by dialysis or gel filtration at the liposome suspension.

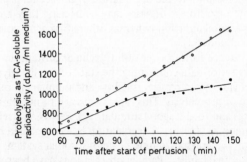

Fig. 4.13 The proteolysis in two liver perfusions is shown. In the upper: (○) Liposomes without pepstatin were added at 105 min (↑); in the lower: (●) Liposomes with pepstatin within were added. The latter regime inhibited proteolysis. See Table 4.4 for a summary of such experiments.

Table 4.4 The inhibition of proteolysis in perfused liver by liposome-entrapped pepstatin

Addition made at 105 min	% Inhibition (Average for groups of at least 5 perfusions)
(a) None	1.0
(b) 2 ml 5 mM potassium phosphate buffer, pH 7.0, containing pepstatin (50 μg ml^{-1})	0.5
(c) 2 ml control liposomes	5.0
(d) 2 ml liposomes containing pepstatin (50 μg ml^{-1})	43.0

are artificial model membranes composed solely of lipid. Pepstatin did not rapidly penetrate the liposomes either and thus could be retained within them. It was known that liposomes could be rapidly phagocytosed by liver cells (in contrast to the slower pinocytosis of free pepstatin by cells). Thus liposome-entrapped pepstatin gained relatively rapid access to liver lysosomes in perfusion, and was found to inhibit proteolysis quite quickly and to a substantial degree (Fig. 4.13, Table 4.4). This constitutes direct evidence for a substantial role of lysosomes in the degradation of intracellular proteins.

Subsequent evidence with the same inhibitor has shown that lysosomes participate in proteolysis of both short and long half-life proteins, and possibly in that of proteins containing amino acid analogues. Non-lysosomal mechanisms are also involved, either in initial reactions which prepare proteins for lysosomal breakdown, or in complete alternative pathways. These are discussed further in Chapter 6.

Evidence of roles of lysosomes on turnover of other materials, in addition to the suggestive evidence from storage diseases, is limited. The phospholipids in lysosomal membranes are known to be replaced at a significant rate, so this may reflect a significant contribution to degradation of phospholipids. But of course, it also reflects membrane transport between plasma membranes and lysosomes and back, and phospholipid exchange via soluble protein carriers.

There is also some indirect evidence that glycosaminoglycans synthesized by cultured fibroblasts are to some degree digested by lysosomes: their degradation is inhibited by chloroquine, and when the cells are cultured in media with raised pH. Both these factors could act by raising intralysosomal pH, and thus reducing proteolysis, and chloroquine might have some directly inhibitory effects on participating enzymes. The hydrolases likely to contribute to lysosomal degradation of the glycosaminoglycan, chondroitin sulphate are shown in Fig. 4.14.

There is no real evidence for degradation of nucleic acids in secondary lysosomes, except that supplied in dead cells (see Chapter 5). But as a already noted, substantial nuclease activity is present in lysosomes. Several intriguing observations suggest that lysosomes can accumulate around nuclei, and that there may be circumstances in which they

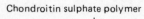
Chondroitin sulphate polymer

1.
            | hyaluronidase

Chondroitin sulphate oligosaccharides

For example:

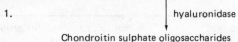
GA—GalNAc—GA—GalNAc—GA—GalNAc
     | | |
    SO_4 SO_4 SO_4

2. | *β-glucuronidase*

Glucuronic acid   +   GalNAc—GA—GalNAc—GA—GalNAc
                    | | |
                   SO_4 SO_4 SO_4

3. | *sulphatase*

Inorganic sulphate   +   GalNAc—GA—GalNAc—GA—GalNAc
                                | |
                              SO_4 SO_4

4. | *β-N-acetylhexosaminidase*

N-Acetylgalactosamine +   GA—GalNAc—GA—GalNAc
                                 | |
                            SO_4 SO_4

Fig. 4.14 Degradation of chondroitin sulphate by lysosomal enzymes.

actually penetrate the nuclear membrane. This has been observed during the response of certain hormone sensitive cells to their triggering hormone; the lysosomes might thus carry hormone to the nucleus, and initiate fresh transcription, perhaps by means of their hydrolases. But this would presumably be an example of limited hydrolysis of the proteins masking the DNA, and would probably not inolve DNA degradation.

The paucity of evidence for lysosomal degradation of several categories of intracellular macromolecule cannot be taken as significant, because as yet there has been little experimental work on their mechanisms of turnover. Indeed, we will see later that the turnover of several categories of cellular material can be most easily explained by invoking a lysosomal route (Chapters 5 and 6). So it is probably fair to assume that that the immense degradative potential of lysosomes is extensively exploited, and that lysosomes are responsible for the breakdown of a substantial portion of most intracellular substrates.

Bibliography

Barrett, A. J and Heath, M. F. (1977), In *Lysosomes: a laboratory handbook* (Ed.) J. T. Dingle, pp. 19–145, North Holland, Amsterdam.
The best review on the subject of lysosomal enzymes.
Dean, R. T. (1975), Direct evidence of the importance of lysosomes in the degradation of intracellular proteins, *Nature (London)*, 414–416.
Dean, R. T. and Barrett, A. J. (1976), 'Lysosomes' *Essays Biochem.*, **12**, 1–40.
A modern review.

Lysosomes in Biology and Pathology, Volumes 1–5 (edited by J. T. Dingle with
    H. B. Fell, [1,2]; alone [3]; with R. T. Dean, [4,5]) North Holland, Amster-
    dam, 1969 (1,2); 1973(3); 1975(4); 1976(5).
The standard source of detailed reviews on all aspects of lysosomal function.
Dean, R. T. (1977), in *Lysosomes; a laboratory handbook*, (Ed.) J. T. Dingle,
    pp. 1–17, North Holland, Amsterdam.
Describes methods.
de Duve, C., (1969), The lysosome in retrospect, In *Lysosomes in Biology and
    Pathology,* Vol. I, (Eds.) J. T. Dingle and H. B. Fell, pp. 3–40, North
    Holland, Amsterdam.
de Duve, C. and Wattiaux, R. (1966), Functions of lysosomes, *Ann. Rev. Physiol.,*
    28, 435–492.
The classic reviews.
Neufeld, E. F., Lim, T. W. and Shapiro, L. J. (1975), Lysosomal Storage Diseases.,
    Annu. Rev. Biochem., **44**, 357–376.
Novikoff, A. B. (1973), Lysosomes: a personal account; In *Lysosomes and Storage
    Diseases,* H. G. Hers, and F. van Hoof, (Eds), pp. 1–41. Academic Press,
    London.
*An account by the co-founder of lysosomology, covering in detail the GERL
concept.*
Vaes, G. (1973), Digestive Capacity of Lysosomes, In *Lysosomes and Storage
    Diseases,* (Eds.) H. G. Hers and F. van Hoof, pp. 43–17, Academic Press.
    London,
A useful description.
Weissmann, G., Dukor, R. and Zurier, R. B., (1971), *Nature, New Biol.,* **231**, 131–
    135. *Lysosomal enzyme secretion by macrophages and polymorphs.*

5 Characteristics of degradation in living cells

This chapter first deals with turnover of macromolecules themselves, and
then proceeds to discuss the turnover of organelles. As an introduction,
Table 5.1 summarizes some of the available data about both these
aspects, concentrating on rat liver, for which most information is available. The most notable feature of this table is the diversity of turnover
rates: and although some organelles seem to turn over as units (so that
all their components show the same half-lives), others possess components
with widely differing half lives.

5.1 Protein degradation

It is really only the proteins which have yet received substantial study as
far as turnover is concerned. So this section outlines general characteristics of *in vivo* turnover, and a subsequent section deals comparatively
with other macromolecules. Many of the general features of protein turnover may have parallels in the turnover of other macromolecules; but this
has only been shown to a limited extent so far.

   (1) Protein turnover is extensive: for instance about 40% of rat liver
protein is degraded every day

Table 5.1 Half-lives for biological macromolecules and organelles

	Approximate half-life (days)
Rat liver components:	
Average protein in homogenate	3.3
Average nuclear protein	5
Soluble (cytosol) proteins (average)	3—5
Fatty acid synthetase (multienzyme complex)	3—5
Average peroxisomal protein	1.5
Plasma membrane proteins	1.5
Average ER protein	2
smooth microsomes	2
rough microsomes	2
NAD glycohydrolase of ER	18
Average mitochondrial protein	4
inner membrane (cytochromes c, b; heme a)	5
outer membrane (cytoch. b_5)	4.5
inducible aminotransferases	less than 1
δ-amino levulinate synthetase (inducible; matrix)	0.1
Peroxisomes (average protein)	2
Lysosomes (average protein)	1
Lysosomal β-glucuronidase	15
Average Ribosomal Protein	5
Ribosomal RNA	5
Mouse liver	
DNA	5
Mouse duodenum	
DNA	1
Rat central nervous system myelin:	
some proteins	35 (?)
average phospholipids	41
Mouse central nervous system myelin:	
different classes of proteins	40—95
Guinea pig pancreas:	
proteins of zymogen granule membranes (average)	4.5
proteins of rough ER membranes (average)	5
Rat costal cartilage polysaccharide	8—16
Rabbit cartilage collagen	more than 50

(2) All proteins so far investigated are degraded.
(3) Hormone levels, drugs, diet and possibly age, can affect the rate of protein degradation. This rate is thus controlled, and can be a major determinant of growth: for instance liver growth in several circum-

Table 5.2 Half-lives of some rat liver enzymes

Enzyme	Half-life
Ornithine decarboxylase (soluble)	10 min
δ-Aminolevulinate synthetase (mitochondria)	60 min
Alanine-aminotransferase (soluble)	0.7–1.0 d
Catalase (peroxisomal)	1.4 d
Tyrosine aminotransferase (soluble)	1.5 h
Tryptophane oxygenase (soluble)	2 h
Glucokinase (soluble)	1.25 d
Arginase (soluble)	4–5 d
Glutamic-alanine transaminase	2–3 d
Lactate dehydrogenase isozyme-5	16 d
Cytochrome c reductase (endoplasmic reticulum)	60–80 h
Cytochrome b_5 (endoplasmic reticulum)	100–200 h
NAD glycohydrolase (endoplasmic reticulum)	16 d
Hydroxymethylglutaryl CoA reductase (endoplasmic reticulum)	2–3 h
Acetyl CoA carboxylase (soluble)	2 d

stances involves a substantial contribution from reduced protein degradation.

(4) The degradation of most proteins is a first order exponential process, which implies that molecules are chosen at random for breakdown (rather than undergoing any 'aging' which renders them more prone to complete breakdown).

(5) Energy is required for some protein breakdown. It seems that one of the chemical reactions involved in the early stages of proteolysis is the most likely candidate for the origin of this requirement: it is observed in bacteria, and thus does not seem likely to be due to the need to transport substrates (for instance, into lysosomes). It also seems that intralysosomal proteolysis, studied when substrates are already within isolated lysosomes, may also show an energy requirement, and such a requirement has also been observed recently in a soluble cell free system. Thus a soluble reaction is probably involved, which requires ATP.

(6) Individual proteins turn over at very different rates. This is indicated in Table 5.2.

(7) Proteins of short half-life are usually: inactivated by heat at a greater rate; degraded more rapidly by lysosomal and other proteinases; more hydrophobic and adsorbed to a greater extent by lysosomal and other hydrophobic membranes and surfaces; composed of larger subunits; more negatively charged (lower isoelectric point); than proteins of long half-life.

(8) Abnormal proteins, such as those containing amino acid analogues, and possibly those synthesized with sequence errors, are degraded faster than normal proteins.

(9) The rate of degradation of several enzymes is decreased when the concentration of their substrates and cofactors are increased.

Although most of these characteristics have been studied primarily for intracellular proteins, they nearly always hold true for extracellular proteins in general. For both classes of protein there are of course specific exceptions (such as the desialylated glycoproteins which are selectively endocytosed by means of receptor molecules on liver parenchymal cells; (see Section 6.1), but the overall similarities suggest that the molecular properties of all proteins are the prime determinant of their turnover rate, and that similar mechanisms are involved in the turnover of both intracellular and extracellular proteins.

5.2 Limited cleavage of proteins

The importance of this phenomenon in the secretion of proteins from some cells, in viral protein maturation and in other processes has already been mentioned. In this section we deal briefly with two major proteolytic cascade systems. In these systems, successive proteolytic cleavages activate successive members of the chain which in turn activate the following members. Complement consists of an array of proteins (C1 to C9) comprising the 'classical' pathway, and a separate array which form an 'alternative' route for certain critical reactions, which then feed into the main route. Not all the activations in complement are proteolytic; and the formation of macromolecular complexes is important also. The processes tend to occur on membrane surfaces.

The complement system of proteins in blood is an important element

First site: activation of recognition unit

1. $S_1 A + Clq \Big<^r_s \rightarrow S_1 A{-}Clq \Big<^r_{\downarrow \bar{s}}$

Second site: assembly of activation Unit

2. $C4 \xrightarrow{S_1 A\overline{Cl}^-} C4a + C4b^*$

3. $C2 \xrightarrow{S_1 A\overline{Cl}^-} C2a^* + C2b$

4. $S_{II} + C4b^* + C2a^* \rightarrow S_{II} \overline{C4b,2a}$

5. $C3 \xrightarrow{S_{II}\overline{C4b,2a}} C3a + C3b^*$

6. $S_{II}\overline{C4b,2a} + C3b^* \rightarrow S_{II}\overline{C4b,2a,3b}$

Third site: assembly of membrane attack mechanism

7. $C5 \xrightarrow{S_{II}\overline{C4b,2a,3b}} C5 + C5b^*$

8. $S_{III} + C5b^* + C6 + C7 + C8 + C9 \rightarrow S_{III}C5b,6,7,8,9$

[a]S_I, S_{II}, S_{III}: topographically distinct sites on target cell surface. A: antibody to cell surface constituent; bar denotes active enzyme; asterisk denotes enzymatically activated, labile binding site.

Fig. 5.1 The classical complement sequence.

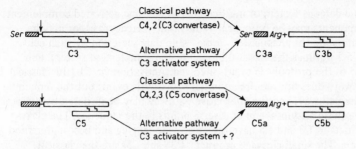

Fig. 5.2 The two pathways for C3 and C5 activation in serum. Arrows indicate sites on C3 and C5 where enzymatic attack causes a release of anaphylotoxins. The C3 and C5 molecules consist of disulphide linked non-identical polypeptide chains.

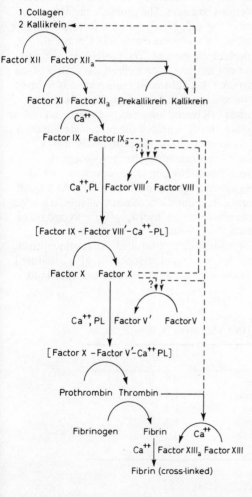

Fig. 5.3 The probable mechanism of blood clotting by the intrinsic system in mammalian plasma.

of the defence system of multicellular organisms: activated complement components may be lytic, or may facilitate phagocytosis, or may be chemotactic for phagocytes, or may stimulate secretion by such cells. Fig. 5.1 outlines the classical branch of the complement system, and some of the proteolytic events involved. Activation via C1 (the classical pathway) does not require an initial proteolytic event; but the remainder of the pathway, and the alternative pathway may be initiated by diverse proteolytic enzymes, and at diverse points in the sequence. The cleavage products of C3 and C5 have the most wide-ranging and best understood effects. The small cleavage products, C3a and C5a are the classical anaphylatoxins which initiate histamine secretion from mast cells; histamine is an important inflammatory mediator, with vasoconstrictive and other effects. (Cb is an inducer of lysosomal enzyme secretion from several cells, as discussed elsewhere. The terminal complex C5b,6,7,8,9 is a lytic complex and the lysis is amongst the classic functions of complement.

Fig. 5.2 illustrates the proteolytic modifications of C3 and C5 under the action of the two complement cascades. The probable mechanism of blood clotting is summarized in Fig. 5.3. Nearly half the factors are serine endopeptidases; and as mentioned already the removal of fibrin during repair is also initiated by a proteolytic cascade. The proteolyses seem to be highly selective; but as yet not much is known about the ability of the component proteinases to attack other substrates. This system is referred to as the 'intrinsic' system, since it is present in serum. There is also an extrinsic system, which involves cell bound enzymes. A further interesting example of limited proteolysis is in the action of the proteinase inhibitor α_2-macroglobulin ($\alpha_2 M$). This seems to 'trap' the proteinases after a limited cleavage which causes a conformational change ('springs the trap'). It thus prevents the proteinases from acting on large substrates such as proteins, though often leaves activity against small substrates. As a result of the reaction, the $\alpha_2 M$-proteinase complex becomes available to a selective endocytosis mechanism in macrophages, liver Kupffer cells and other phagocytes. The complexes are thus cleared from circulation, and proteinase activity in serum is thus closely controlled. After endocytosis, both components of the complex are degraded; luckily $\alpha_2 M$ is denatured under lysosomal conditions (acid pH) and thus becomes susceptible to complete breakdown.

Table 5.3 DNA Half-lives in mouse tissues

Tissue	Half-life (days)
Duodenum	1
Bone marrow	1.1
Thymus	1.3
Spleen	1.6
Lymph node	3.1
Liver	4.8

5.3 Degradation in vivo of other macromolecules

Whereas turnover of proteins is mainly occurring in living cells, it seems that a significant proportion of DNA degradation is associated with breakdown of cells following cell death (see Section 5.5.10.). Thus the more rapidly turning over tissues tend to have more rapid turnovers of DNA. This is illustrated in Table 5.3.

There seems to be a little DNA in cytoplasm in addition to the bulk in nuclei, and this may undergo intracellular turnover. Similarly the DNA of mitochondria and chloroplasts, much of which is in the form of circular DNA, probably undergoes some digestion *in situ* but is mainly degraded when mitochondria are removed by autophagy. There is evidence for the occurence of mitochondrial DNAse activity. There is no detailed information about selective degradation of different DNA molecules, but there do seem to be physiological controls on degradation.

A little more is known about RNA turnover. For instance, although the half-life of bulk mRNA is long (several days), and globin mRNA in reticulocytes (which mature to red blood cells, and which cannot make RNA, since they are anucleate) seems also to have a long half-life, tyrosine aminotransferase mRNA in hepatoma cells seems only to have a half-life of about 2h. However very few measurements of turnover have been direct, in the sense of quantitating retention of radioactivity in defined RNAs (see Chapter 2), and so they are not free of ambiguity.

Recently it has become apparent that some eukaryotic mRNAs contain untranslated sequences at both ends. The 'cap' at the 5' end is present in most messages and seems to be involved in the initiation of protein synthesis. At the 3' end of some mRNAs is a sequence of adenylic residues, which is added on to the mRNA after transcription. This poly(A) sequence progressively shortens during ageing of the molecule in the cytoplasm. Since the poly (A) seems to be found on messages with long half-life (such as that for globin) and not on those of short half-life (e.g. histone mRNA) it was suspected that the poly(A) tail was somehow responsible for the stability of these mRNAs. This was studied directly by injecting globin mRNA into oocytes of the frog *Xenopus* and following its stability in comparison to that of mRNA from which the poly(A) had been removed enzymatically. Native molecules were found to be stable while poly(A)-free ones were unstable, thus confirming the suggestion. Of course, other features may also be involved in the stability of mRNAs. In contrast ribosomal RNAs (rRNAs) seem to be degraded with the same half-lives as ribosomal protein; the ribosome may thus be degraded as a unit (see Table 5.1 and Section 5.5.8).

Some general statements can also be made about the turnover of proteoglycans. Values of 8–16 days have been obtained for the half-lives of connective tissue proteoglycans; and it has been shown that the protein and polysaccharide portions have the same half-life, and so presumably are degraded together. Of course this is what one would expect if the total breakdown occurs after endocytosis, in lysosomes. There is also a correlation between the solubility (or extractability) and metabolic turnover; the more soluble molecules turn over more rapidly.

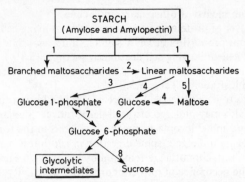

Fig. 5.4 Enzymes in the degradation of starch. (1) α-amylase (maltosaccharides refer to a range of oligosaccharides) (2) Limit dextrinase (3) Phosphorylase (4) α-Glucosidase (5) β-Amylase (6) Hexokinase (7) Phosphoglucomatase (8) Sucrose phosphate synthetase and phosphatase.

There is in addition to intracellular degradation of proteoglycan after endocytosis by liver and connective tissue cells, an intracellular pool in cells which synthesize proteoglycans (fibroblasts, connective tissue cells) in which newly synthesized molecules are degraded (with a half life of about 7 hours).

Starch is a major storage material in many higher plants; and although many of the enzymes which can interact in its catabolism are characterized, the control of their interaction is less well understood, than that of the major animal storage polysaccharide, glycogen (see Section 6.2). Figure 5.4 illustrates the known pathways in starch degradation.

The turnover of phospholipids is now beginning to receive considerable attention. Some early data indicating the diversity of turnover are shown below (Table 5.4). Diverse rates for phospholipids have also been reported for cells in culture; and it is in such systems that the study of control of turnover has reached some substantial conclusions (see Section 6.2).

Extracellular lipid catabolism is also important, particularly in the supply of lipids by liver to other tissues. Transport in plasma depends on the incorporation of the apolar lipids into soluble lipoproteins of various

Table 5.4 Lipid half-lives in myelin sheaths of nerve and in mitochondrial membranes

Lipid	Half-life	
	Myelin	*Mitochondria*
Cerebroside	13 months	2 months
Sphingomyelin	10 months	1 month
Phosphatidyl choline	2 months	2 weeks
Phosphatidyl ethanolamine	7 months	4 weeks
Phosphatidyl serine	4 months	3 weeks
Phosphatidyl inositol	1.25 months	2 days

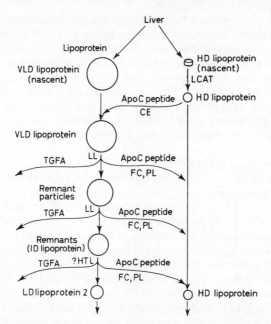

Fig. 5.5 Intravascular lipoprotein metabolism.
Nascent VLD and HD lipoproteins are converted into their mature form
after secretion by the liver. VLD lipoprotein is partially delipidated by
lipoprotein lipase (L.L.), yielding a series of remnant particles. A hepatic
triacylglycerol lipase (H.T.L.) may participate in the later stages of
delipidation. The end product of these processes is LD lipoprotein 2. HD
lipoprotein is transformed into its mature configuration during esterific-
ation of its free cholesterol by lecithin—cholesterol acyltransferase
(L.C.A.T.). Abbreviations: CE — cholesteryl ester; FC — free cholesterol;
PL — phospholipid; TGFA — triacylglycerol fatty acids.

kinds. These constitute a spectrum of particles, usually classified on the
basis of their behaviour during separation procedures in the ultracentri-
fuge. Intravascular lipoprotein metabolism is summarized in Fig. 5.5. The
catabolism of the end-product Low-density lipoprotein (LDL) seems to
be an intracellular event, and it is discussed later (Section 6.2).

5.4 Limited degradation of macromolecules other than proteins

Several examples of limited degradation which lead to endocytosis and
complete breakdown have been mentioned. But this section concentrates
on limited cleavages in the processing of nucleic acid. This is a vast
subject, and can only be given very superficial treatment here.

In the case of DNA there are two main categories of limited cleavage:
those in repair and those in restriction. In the former, damaged segments
of DNA containing non-complementary base pairs in the twin strands are
excised by endonucleases, and repaired by ligases. Much of the work has
been on ultra-violet induced photoproducts, such as pyrimidine-
pyrimidine dimers.

Restriction enzymes are bacterial endo-deoxyribonucleases. Each restriction enzyme requires a recognition site of 4—8 base pairs; some cleave at the recognition site while others cleave at some fairly defined distance from it. The enzyme is also associated with modifying activities which protect the endogenous DNA by specific nucleotide methylation at the recognition site. Thus foreign DNA, such as that from a newly invading bacteriophage, is unmodified and so susceptible to restriction. Established phage DNA on the other hand, has become modified and so resistant. Restriction enzymes have not yet been found in eukaryotes.

RNA processing is an extremely complex subject. But it is clear that both in eukaryotes and in prokaryotes normal rRNA molecules are formed by limited cleavage of much larger precursors. In mammalian cells rRNA is made as a large molecule (characterized by its sedimentation constant as 45S) in the nucleoli within the nucleus. Cleavage is performed by an endonuclease, and the terminals of the precursor are also trimmed by exonucleases. The 45S component is converted via a 32S one, to the 28S and 18S molecules found in rRNA.

Similarly mRNA molecules represent a small fraction of a large precursor nuclear RNA termed hnRNA (heterogeneous nuclear RNA). The insertion of poly(A) which has already been discussed seems to take place place on the hnRNA, which then undergoes extensive degradation, with the exception of the material which contains poly(A). Free poly(A) is known to be a good inhibitor of several endonucleases, and it is thought that it may function as such within the heterogenous nRNA. It would at the same time itself undergo slow digestion, so that the eventual catabolism of the mRNAs could be ensured.

5.5 Turnover of organelles and cells

5.5.1 Plasma membrane

There seem to be two main categories of plasma membrane turnover. In one category, most of the membrane proteins show very similar half-lives, and some of the other membrane components show similar half-lives. This 'unit degradation' is most easily explained as non-selective endocytosis of membrane carrying it to the lysosomal system for complete degradation. That this is the fate of much endocytosed membrane protein has been clearly shown in work on macrophages and cultured 'L-cells'. While the similarity of protein half-lives is so great as to indicate that most breakdown occurs in this way, the other components may undergo significant degradation *in situ*, as membrane glycosidases and phospholipases are known. This local degradation will contribute heterogeneity to the turnover of the phospholipids, as is usually observed, and to other components. As mentioned earlier, phospholipid exchange between membranes, mediated by soluble protein carriers, makes the interpretation of turnover of lipid in a particular membrane very difficult.

The second type of plasma membrane turnover shows heterogenity of all components' half-lives, including protein. This occurs in many mammalian cells, and probably in most bacterial plasma membranes. The

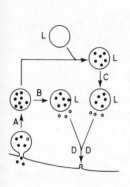

Fig. 5.6 Mechanisms of plasma membrane recycling. After endocytosis (A) of soluble materials ⊙, the endocytic vesicle may divide to form one large and several smaller vesicles (B). Alternatively, a similar vesicle division may occur after the endocytic vesicle has fused (C) with a lysosomes (carrying lysosomal enzymes, L). In both cases, the soluble materials, are largely retained in the large vesicle. The small vesicles may then fuse with the plasma membrane (D), thus recycling membrane.

mechanism is not well understood, but is discussed in Section 6.1.

A common feature of plasma membrane turnover in endocytic cells is that membrane is internalized by endocytosis much more rapidly (up to 10 times) than it is degraded. This shows that not all the endocytosed membrane is degraded by lysosomes; and there is now evidence that much is returned to the plasma membrane ('recycling'). Two possible routes of recycling are shown in Fig. 5.6. In one, a small vesicle leaves the endocytic vesicle (before it has become a lysosome) and fuses again with the plasma membrane. In the other, the same process occurs after the endocytic vesicle has joined the lysosomal system by fusion, becoming a secondary lysosome. Both mechanisms supply membrane back to the plasma membrane and thus reduce its rate of degradation. The soluble contents of the endocytic vesicles seem to be selectively retained within the remaining endocytic vesicle (because tracer enzymes which have undergone endocytosis are not re-exocytosed at a rate comparable with the rate of recycling). How this selective retention is achieved is not known.

5.5.2 Cytosol

Cytosol components show very diverse half-lives. The protiens show most of the general characteristics listed earlier (correlations of turnover with subunit size and molecular charge of the intact polymer). As pointed out earlier, this implies that the structure of a protein is an important determinant of its turnover rate. But it is notable that these relationships do not hold when animals are starved or diabetic. This may simply be because of the increased activity of a rather non-selective mechanism of degradation which will have a large effect on the rate of turnover of long half-life proteins, but only a small effect on that of short half-life proteins. Thus it will make turnover rates much more homogeneous, and obscure the molecular relationships.

5.5.3 Endoplasmic reticulum (ER)

ER proteins turnover rather like those of the second type of plasma membrane, showing heterogeneous turnover. But unlike the latter proteins,

ER proteins show good correlations between sub-unit size and half-life, as for cytosol proteins.

ER is the site of extensive phospholipid synthesis and also contains phospholipases. Phospholipid turnover is heterogeneous, and presumably includes export of newly synthesised molecules, tranport to lysosomes via autophagy, and *in situ* breakdown.

5.5.4 Mitochondria

The component proteins of the inner mitochondrial membrane in eukcaryotes show quite similar half-lives and thus this membrane may like some plasma membranes undergo unit degradation. The turnover of mitochondrial nucleic acid seems to have a similar kinetic, and thus may occur in the same way. On the other hand, some other proteins and other components show very heterogeneous turnover, so other mechanisms are also involved.

5.5.5 Lysosomes

Turnover of lysosomal constituents is also heterogeneous. But interestingly, the proteins of neither the soluble nor the membrane phase show the subunit size—turnover correlation; and lysosomal glycoproteins apparently lose their carbohydrate side-chains during their intralysosomal existence.

It is quite possible that there is an interesting intralysosomal recycling mechanism, which minimises the expenditure of lysosomal membrane during lysosomal autophagy. This would have very similar characteristics to plasma membrane recycling, in particular soluble autophaged materials could be retained inside the lysosome even while some lysosomal membrane is recovered. This mechanism is illustrated in Fig. 5.7.

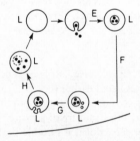

Fig. 5.7 Lysosomal membrane recycling: the lysosome may invaginate to take in material from the cytoplasm (E). This material is at first in an intralysomal vesicle, and so separated by a membrane from the lysosomal enzyme enzymes (L), and at this stage the vesicle may divide (F), in a manner analogous to the division in plasma membrane recycling (B). The small vesicles formed may rejoin the lysosomal membrane by fusion (G), while most internalised material is retained within the lysosome, and after disintegration of the intra-lysosomal membrane (H) becomes accessible to lysosomal enzymes and undergoes degradation.

5.5.6 Peroxisomes

Although the turnover of separate peroxisomal components has not been studied much, the turnover of the organelle as a whole has been followed. They seem to be degraded at random; each particular peroxisome seems to have the same chance of undergoing breakdown. It is possible that all peroxisomes form part of a single interconnected system which of course would explain this finding about their turnover!

5.5.7 Secretory vesicles

In certain hormone secreting cells in the pituitary excess secretory vesicle can be removed by fusion with lysosomes (Fig. 4.5). This is a special form of autophagy (since intracellular material enters the lysosomal system) and is termed 'crinophagy'. It has only been studied morphologically but does seem to be accentuated under conditions in which secretion is inhibited experimentally. Thus it may be a controlled biological reaction, which regulates the output of secretory proteins. A related process may be involved in the degradation of newly synthesized proteoglycans within fibroblasts (Section 5.3); so crinophagy may be operative in other systems, such as serum protein secretion from liver parenchymal cells.

5.5.8 Ribosomes

In many types of cell ribosomes seem to undergo unit degradation: protein and rRNA half-lives are similar.

Table 5.5 Modes of destruction of cells during physiological and pathological degeneration. (Modified from Locksin and Beaulaton, (1974) *Life Sci.*, **15**, pp. 1549–1565.

System	Relative importance of	
	Intracellular lysosomal activity including autophagy	Activity of phagocytic cells in endocytosing large cell debris
Embryonic		
Sympathetic nerve ganglia	0	+
Developmental and hormonal		
Arthropod muscles and nerves	++	0
Tadpole tail	++	+++
Some annelid tissues	0	+++
Pathological		
Lethal anoxia	0	+++

Rough quantitation is on a scale from 0 (little importance) to +++ (very great importance).

5.5.9 Intracellular degradation during erythrocyte maturation

In mammals erythrocytes mature by losing intracellular organelles including nuclei. While most of the organelles seem to be autophago-cytosed and hence degraded, the nuclei seem to be extruded. Their fate is then like that of dead and shrunk cells described next: phagocytosis.

5.5.10 Degradation of cells during normal maintenance of organs and in developmental processes.

Normal cell death in tissues often occurs by a process of 'shrinkage necrosis' (which has been termed ('apoptosis'). It is not initiated by lysosomal action within the dying cell except under pathological conditions of rare incidence. But the dead cells are usually endocytosed by phagocytes and degraded by their lysosomes. Many cells have quite short lifespans, and there is considerable variation in this parameter from cell type to cell type.

In situations of remodelling, or of pathological regression, there may be an induction of autophagy in the regressing cells, but there is usually an invasion of macrophages which remove the cells in the same way as they do for normal cell death. Table 5.5 lists several situations in which there is physiological cell death, and indicates the degree of involvement of internal lysosomal activity and of invading phagocytes. In these cases too, there is no evidence that the internal lysosomal activity is an initiator: rather it is merely a participant.

Bibliography

Arber, W. (1977), 'What is the function of DNA restriction enzymes?', *Trends Biochem. Sci.* 2, N176–N187.

Ballard, F. J. *Essays in Biochemistry,* 1977 13, 1–37.
Protein degradation.

Capecchi, N. R., Capecchi, N. E., Hughs, F. M. and Wahl, G. M., (1974), *Proc. Natn. Acad. Sci.,* U.S.A., 71, 4732–4736.
Degradation of proteins containing amino acid analogues.

Davie, E. W. and Fujikawa, K. (1975), Basic mechanisms in blood coagulation, *Annu. Rev. Biochem.* 44, 799–829.

Goldberg, A. L. and St. John, A. C. (1976), Intracellular protein degradation in mammalian and bacterial cells, *Annu. Rev. Biochem.,* 45, 747–803.
An excellent review.

Grossman, L., Braun, A., Feldberg, R. and Mahler, I. (1975), Enzymatic repair of DNA, *Ann. Rev. Biochem.,* 44, 19–43.

Kerr, J. F. R. (1973), Some lysosome functions in liver cells reacting to sublethal injury, In '*Lysosomes in Biology and Pathology,* Vol. 3, (ed.), Dingle, J. T. pp. 365–394: North Holland, Amsterdam.
Covers shrinkage necrosis.

Lewis, B. (1977), Plasma lipoprotein interrelationships, *Biochem. Soc. Trans.* 5, 589–601.
A useful concise review.

Manners, D. J. (1974), *The Structure and Metabolism of Starch, Essays Biochem.* 10, 37–71.

Muller-Eberhard, H. J. (1975), *Complement, Annu. Rev. Biochem.,* 44, 697–724.

Perry, R. P. (1976), Processing of RNA, *Ann. Rev. Biochem.,* 45, 605–629.

6 Mechanisms and control of cellular degradation

6.1 Mechanisms for selectivity of degradation

We have already seen that degradative enzymes are widespread in cells, and that eukaryotic cells have a specialized degradative organelle, the lysosome. Now we consider how the activities of the various degradative routes are integrated, and how they work with the selectivity which is observed. And it should be admitted immediately that we here deal more with hypotheses than with established mechanism.

We have already noted that in multicellular organisms, the turnover characteristics of extracellular soluble proteins and intracellular proteins are surprisingly similar. Since the degradation of both types of protein occurs intracellularly, and since lysosomes are involved to at least some degree in both events, it seems possible that a similar step controls selectivity of both processes. This may be binding to membrane surfaces. Proteins interact to various degrees with membranes, and as Jacques pointed out (Fig. 6.1), this could allow selective endocytosis of proteins: those proteins which bind well are taken up proportionally faster than those which do not bind. As a spectrum of binding affinity can be envisaged, depending on gross molecular parameters such as surface area (depending on size, hydrophobicity etc), a similar spectrum of uptake rates could result. Since uptake seems inevitably to result in degradation,

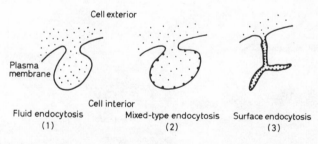

Cell exterior

Plasma membrane

Cell interior

Fluid endocytosis (1)

Mixed-type endocytosis (2)

Surface endocytosis (3)

Fig. 6.1 Selectivity in pinocytosis (after Jacques in *Lysosomes in Biology and Pathology*, Vol. 1: see Bibliography to Chapter 3). Molecules may enter either in solution (1) or bound to the surface (3), or by a combination (2). Process (3) can allow entry of a much larger proportion of external molecules per unit time than process (1), and the proportion in (3) varies from molecule to molecule, according to the degree to which the molecule binds to the plasma membrane. In contrast all molecules entering a cell by process (1) enter at the same proportional rate, i.e. non-selectively. Thus, the combination of processes 1–3 acting on a mixture of molecules allows considerable diversity in proportional entry rates.

degradation rates would to a fair approximation reflect uptake rates. Experimentally it does seem that rapidly turning-over proteins undergo more rapid endocytosis by cells than slowly turning-over ones.

We should note at this point that the parameters of the molecular structure of proteins which determine their turnover thus seem to dictate their membrane binding avidity in a similar sense. No specific receptor interactions are postulated; and similar gross selectivity on the basis of physical properties of the macromolecules seems also to apply in the endocytosis of other macromolecules, such as proteoglycans: amongst which dermatan sulphate chains are endocytosed the most avidly.

Specific receptors are also involved in the catabolism of extracellular molecules; the best studied are the several distinct receptors for at least three different saccharides on the terminal position of glycoprotein side chains. Of these receptors the first characterized, on rat liver parenchymal cells, is specific for galactose termini on de-sialylated glycoproteins. Rapid endocytosis and degradation ensues from interaction with the receptor, and the plasma survival of the desialylated version of most glycoproteins is only a matter of minutes while that of the native molecule may be up to several hours. This is thus a highly selective system.

The other well characterized receptor system is that for circulating low density lipoprotein (LDL) on fibroblasts (Fig. 6.2). It is similar in function to that on liver cells, in that it allows selective endocytosis and degradation. But it is also known to have two additional features: the product of degradation, cholesterol, suppresses the activity of the cells endogenous cholesterol-synthesizing machinery, thus avoiding overproduction of cholesterol in the cells; and the LDL receptor concen-

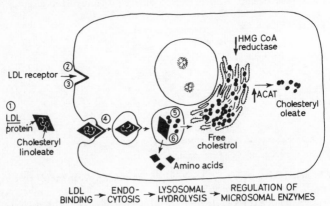

Fig. 6.2 Sequential steps in the LDL pathway in cultured human fibroblasts. The numbers indicate the sites at which mutations have been identified: (1) abetalipoproteinemia: (2) familial hypercholesterolemia, receptor-negative; (3) familial hypercholesterolemia, receptor-defective; (4) familial hypercholesterolemia, internalization defect; (5) Wolman syndrome; and (6) cholesteryl ester storage disease. HMG-CoA reductase denotes 3-hydroxy-3-methylglutaryl coenzyme-A reductase, and ACAT denotes acyl-coenzyme A: cholesterol acyltransferase.

tration is regulated in response to the external concentration of LDL. When LDL is present at high levels, the receptor number decreases to a value only 10% of the maximum number found when little LDL is available. This regulation ensures that the uptake of LDL is just sufficient for cellular requirements for cholesterol.

Non-receptor mediated selective binding of protein to membranes may be important in the degradation of intracellular proteins as well as for extracellular proteins. For the same kind of selectivity of binding may occur on the surface of membrane which contribute to autophagy (ER, lysosomes), and thus lead to selective uptake of proteins into the lysosomal catabolic pool, with resultant selective degradation.

On the other hand, the proteolytic susceptibilities of proteins also correlate very well with their turnover rates. So any proteolysis of freely soluble molecules in the solution containing them could result in the observed specificity. This may be important in the turnover of cytosol proteins and of many bacterial proteins (since bacteria do not have lysosomes). In addition while unit degradation of membranes and organelles is strongly suggestive of lysosomal action, the superimposed heterogeneity may reflect either free digestion *in situ* or some exchange process whereby molecules gain access to a soluble degradation system, such as may occur in the cytosol. It is noteworthy that proteolysis of proteins *in situ* on membranes can have quite different characteristics from proteolysis in solution. For whereas amongst soluble proteins, those of large subunit are degraded most rapidly, amongst membrane bound proteins, the converse may be true. This could explain the lack of subunit size–turnover correlation in lysosomal membranes if they are substantially degraded *in situ*. The lack of correlation for the soluble lysosomal proteins is more obscure: though it may depend on reversible interactions between the proteins and the membrane.

Since the correlation of turnover and molecular size is closest related with sub-unit size, it is logical to suppose that the real substrate for the degradative machinery is the sub-unit. This view is supported by the observation that different sub-units of heteropolymeric proteins (composed of more than one type of sub-unit) often turn over at quite different rates. The dissociation into subunits may involve the exposure of increased hydrophobic areas (capable of binding to substrate) or merely of more proteinase sites. Besides dissociation, there may be other initiating reactions in proteolysis, catalysed by limited cleavages, or by modification of disulphide bridges with consequent structural rearrangement, or whatever.

A fascinating possibility which has emerged recently, is that a soluble calcium-activated proteinase, which has been found in several cell types may be concerned with the initiation of degradation of some proteins. This might form a system which could be controlled by calcium levels in the cell, which are clearly capable of precise control in view of their role in cell activation discussed elsewhere.

An alternative set of initiating agents are the putative group-specific proteinases. These may be capable of inactivating groups of enzymes

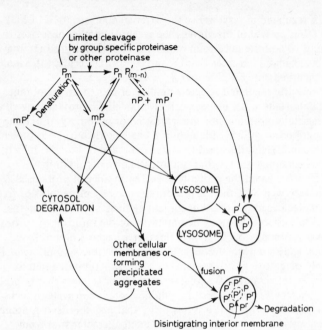

Fig. 6.3 Possible mechansims in the degradation of soluble proteins in cells: P – polypeptide chain; P^1 – polypeptide after limited cleavage; P^r – polypeptide in random coil configuration resulting from denaturation. The quantitative distribution of molecules between the various states of association, denaturation and cleavage is likely to vary from protein to protein.

sharing certain common mechanistic features (e.g. dependence on pyridoxal phosphate), by limited cleavage. These cleavages might prepare the proteins for complete digestion. Evidence on the existence of such enzymes is quite limited, and as yet, not entirely satisfactory. Evidence on their role is negligible. Fig. 6.3 summarizes several possible routes of protein degradation, with reference to the eukaryotic cell. But while the prokaryotic cell lacks lysosomes, it tends to have aggregates of degrading proteins, particularly obvious when cells have incorporated amino acid analogues. These aggregates are sedimentable, and thus precipitation or adsorption to the aggregate may serve the same sort of functions that are postulated for lysosomes. Molecular parameters may influence aggregation in the same way they do adsorption. This is an area of intense investigation at present, but no firm conclusions have yet been reached.

We have already noted that structural features of other macromolecules may affect their turnover rates (such as the poly(A) tails of mRNAs). But with the exception of the suggested mechanism by which poly(A) could inhibit RNAse activity and thereby prolong survival of mRNAs, there is really little idea *how* such features of molecules affect their turnover. Mechanisms analogous to those discussed for proteins may well be feasible.

6.2 On the control of turnover

It is known that control of degradation rate plays a substantial role in changes of levels of certain proteins in response to hormonal signals. Similarly, changes of average protein turnover are important in many physiological conditions. Degradation is often accelerated during starvation of animals; this provides the nutrients which are no longer available from food. Similar responses are observed when cultured cells are 'starved' of serum. This observation holds for prokaryotes and eukaryotes: but it has been suggested that in the latter the only involvement of lysosomes in degradation of protein is in the stimulated degradation in response to nutritional deprivation. Since the acceleration can occur in cells without lysosomes when they are starved it is clearly possible that several different mechanisms are involved in the acceleration. And the evidence from inhibitor studies, and from the characteristics of turnover in perfused organs quite strongly implicates lysosomes in both basal (normal) and accelerated (in starvation) turnover.

Although accelerated protein degradation normally accompanies starvation, and degradation is reduced in situations of rapid growth, there are cases where rapid growth involves *increased* degradation, accompanied by even more greatly increased protein synthesis. It seems that in transformed (cancer-like) cells in culture, where growth is accelerated in an uncontrolled manner, protein degradation is depressed, rather as it is in normal rapidly growing cells. But it is more sensitive to regulation by the hormone insulin, which is a general inhibitor of protein turnover in eukaryotic cells. In this sense the proteolysis in these cells shows abnormal control responses too.

Besides the hormonal effects of insulin (inhibitory) and of glucagon (accelerating), the main controlling factor in general degradation seems to be amino acid supply. This is depressed in starvation where rapid breakdown of protein obtains, and elevated to normal in well fed animals, where lower rates obtain. Thus it seems to be an inhibitor; and this has been confirmed in many experimental situations. Probably a crucial factor in determining the ongoing rate of proteolysis is the level of guanosine tetraphosphate. In both eukaryotic and prokaryotic cells, this metabolite varies in response to starvation (or nutritional 'step down' in cultured cells and bacteria) and many provide a direct signal to the proteolytic system.

The mechanism of inhibition or stimulation by the hormones and amino acids are not understood. But there is one area in which the mechanism of regulation by amino acid supply is at least suspected. In protein deficiency in rats, the turnover of both ribosomal protein and rRNA in liver is accelerated in parallel. The initial response to the low protein diet involves an accumulation of ribosomal subunits in the cytosol. This may result from the reduced availability of amino acids for protein synthesis, and thus the reduced requirement for membrane-bound ribosomes. It is thought that protein deficiency is accompanied by a decrease in the level of a cytosol RNAse inhibitor, and thus by an increased expression of RNAse activity. This could facilitate the increased turnover of rRNA. But

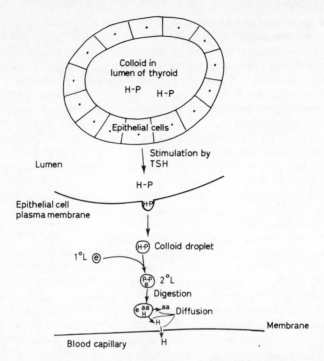

Fig. 6.4 Lysosomal processing in secretion of thyroid hormone.
The hormone (H) is initially released into the thyroid lumen, in colloidal
form as a complex with a large protein (H-P; thyroglobulin) Following
stimulation by thyroid stimulating hormone (TSH), accelerated endocytosis of H-P occurs. The endocytic vesicle fused with primary lysosomes, and degradation of the protein moiety releases free amino acids,
and also the low molecular weight hormone (thyroxin). The hormone can
then diffuse into the nearest blood capillary, and enter the circulation.
Within membranes: aa — amino acids released by digestion; e — lysosomal
enzymes. *Outside membranes:* $1°L$ — primary lysosome; $2°L$ — secondary
lysosome.

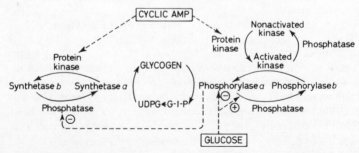

Fig. 6.5 The control of glycogen metabolism in liver. Inhibitory
influences are shown as ⊖ , and stimulation ones as ⊕ .

of course the mechanism of increased proteolysis of the ribosomes, and of the loss of RNAse inhibitor, are not understood. Since the availability of ribosomes seems to be a limiting factor in determining the rate of protein synthesis, increased breakdown of ribosomes leads to decreased protein synthesis during the early response to low protein diets. This illustrates a control exerted by proteolysis on protein synthesis.

At least one example of a limited proteolysis whose rate is controlled is known. This is in secretion of thyroid hormone. Here a large precursor is stored in the lumen of the gland, and when the gland is stimulated by thyroid stimulating hormone, the precursor is endocytosed. Extensive degradation of the protein then produces the small thyroxin molecule, which is secreted (see Fig. 6.4.).

Turning to the other classes of macromolecules, we again meet a paucity of information. In cancer cells it is suspected that there may often be increased RNAse activity, both in lysosomes and cytosol. And it has been observed that these levels return to normal during regression of mammary gland tumours. But again the significance of these observations is obscure.

Quite a substantial amount is known about the control of the glycogen metabolism in liver. This storage compound can be rapidly mobilised, thus fulfilling its role as a store. Some of the complicated interactions between active (phosphosylase a and synthetase a) and in-active (b) forms of degrading and synthesizing enzymes, largely controlled by protein kinases which phosphorylate the protein, are shown in Fig. 6.5. It is notable that glucose, the product of degradation, is a regulator. The figure does not illustrate lysosomal degradation; this is thought to occur by autophagy, and may well be at quite a significant rate, since there are lysosomal storage diseases in which glycogen accumulates.

Phospholipid catabolism has recently become a focus of interest, because, many activated cells (triggered by ligands, hormones etc.) show elevated phospholipid catabolism. Table 6.1 lists some of the cells which are known to show changes in phospholipid turnover when stimulated. In many of these studies, catabolism has not been directly measured, but such measurements as are available support the view that the changes are

Table 6.1 Cells showing alteration of phospholipid turnover when stimulated in culture

Cell	Stimulant	Phospholipid stimulated
Lymphocytes	Mitogens	All phospholipids
3T3 (fibroblasts)	Serum	All phospholipids
Lymphocytes	Phytohaemagglutinin	Phosphatidylinositol
Mouse embryo fibroblasts	Low population density	Phosphatidylinositol and phosphatidylethanolamine

Table 6.2 Phosphatidyl inositol turnover in cells responding to extracellular stimuli

Stimulus	Tissue(s)	Phosphatidylinositol turnover
Muscarinic cholinergic	Various	↑
α-Adrenergic	Various	↑
Serum factors	Fibroblasts in tissue culture	↑
Phytohaemagglutinin and other mitogens	Lymphocytes (T?)	↑
Insulin	Adipose tissue	↑
Glutamate	Insect flight muscle, nervous tissue	↑
ADP, thrombin collagen, adrenaline	Platelets	↑
Phagocytosable material	Polymorphonuclear leucocytes	↑
Glucose	Islets of Langerhans	↑
High K^+ (depolarising)	Nervous tissue, vas deferens	↑
Thyroid-stimulating hormone	Thyroid gland	↑

quite widespread. The listed responses are associated with cell division; and it has frequently been found that elevated phospholipid synthesis is concomitant with cell division. But it has also been found that phosphatidyl inositol turnover is vastly accelerated within seconds of stimulating many responsive cells with appropriate ligands. This is true of lymphocytes (as listed above) and of many types of secretory cell (see Table 6.2), in the absence of cell division.

As discussed early calcium influx seems to be the direct trigger for the cellular responses in many cases. In some systems the elevated phosphatidyl inositol (PI) turnover can be bypassed by means of the calcium ionophore, A23187, which makes available a pathway for calcium through the cell membrane. A23187 was used to induce mast cell secretion of histamine, in the absence of any normal stimuli, and has subsequently been found effective in many secretory and excitable cells. Such experiments confirm that calcium levels in the cytosol are central in the induction of response: and it has been noted that in some systems, calcium is supplied not from the exterior of the cell but from intracellular stores. In view of the occasional disparity between the induction of calcium flux, and the elevation of PI turnover in these experiments with the ionophore, the relationship between the two is not quite clear. But because of the great rapidity of response of PI turnover in many cells (see Table 6.3) it still remains a possibility that with normal stimuli, the PI turnover precedes, and is prerequisite for, the calcium influx and subsequent response. PI turnover may actually directly result in the opening of a calcium gate: such possibilities are illustrated in Fig. 6.6, which also

Table 6.3 Timing of responses in phosholipid metabolism in stimulated cells

Cell	Stimulus	Earliest response
Platelets	ADP	2 s: 3 min
Thyroid	TSH	2 min
Lymphocytes	Phytohaemagglutinin	3 min
Pancreas	Pancreozymin	$\approx$ 5 min
	Acetylcholine	$\approx$ 5 min
Cerebral cortex	Acetylcholine	10 min
	Noradrenaline or phenylphrine	5 min
Parotid gland	Acetylcholine	5 min
Fibroblasts	Serum factors	5 min

includes the notion that Ca^{2+} may be required for the elevated PI turnover, which then allows Ca^{2+} entry. This is consistent with the observation of a calcium-sensitive PI degrading enzyme in some cells. PI turnover may have effects on the activity of membrane adenyl cyclase; and it is well known that changes in the intracellular cyclic nucleotides (cAMP, cGMP, etc) occur during the early stages of cellular responses to triggers; probably constituting important second messengers. The figure also indicates the possible effects of the calcium flux on cGMP levels, via activation of a calcium-sensitive guanyl cyclase.

Thus phospholipid turnover is central in cell activation. It is also quite possible that limited proteolysis is involved, since certain proteinases can activate cells. In some of these cases, it has been shown that limited proteolysis of the cell surface occurs, often resulting in the modification or disappearance of a very few polypeptides. Proteinase inhibitors are often effective in blocking the proteolysis and the activation. It may be that proteolysis allows protein clustering in the membrane to form a calcium channel; or it may interact in some way with the PI turnover system

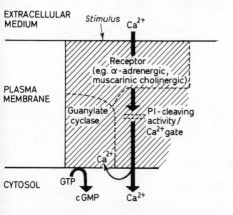

Fig. 6.6 Possible relationships between phosphatidyl inositol breakdown, calcium movements and cell activation.

just described. In a related way, phospholipid turnover may provide increased amounts of precursors (such as arachidonic acid) for the synthesis of prostaglandins. These compounds are known to have many stimulatory and regulatory effects on cells, and so it can easily be imagined that generation might be involved in a switching off of the cellular response, or conversely, an augmentation of it.

It is notable that elevated turnover is concomitant with many events in which membrane fusion takes place. Besides those already mentioned (secretion and cell division) one might point out phagocytosis by macrophages and neutrophils. So it may have some significance in allowing fusion, for it is known that membrane constitution of cells about to fuse is often changed to a more 'fluid' array, and that the membrane of newly formed phagosomes is more fluid than that in the cell surface from which it derives.

It should be apparent from this brief survey that degradative processes have important control functions both in dictating the levels of macromolecules in cells, and outside them, and in the initiation and maintenance of many more complex integrated cell functions. The relative paucity of information about components other than proteins probably conceals an absolute myriad of regulatory functions!

Bibliography

Dean, R. T., (1975), *Biochem. Biophys. Res. Commun,* **67**, 604–609.
On selectivity in autophagy.
Goldstein, J. L. and Brown, M. S. (1977), The low density lipoprotein pathway and its relation to atherosclerosis, *Ann. Rev. Biochem.,* **46**, 897–930.
Excellent on LDL uptake.
Gregoriadis, G. (1975), The catabolism of glycoproteins, In *Lysosomes in Biology and Pathology,* Vol. 4, (Eds.) J. T. Dingle and R. T. Dean, pp. 265–294, North Holland, Amsterdam.
Covers the liver uptake system.
Michell, R. H. (1975), Inositol phospholipids and cell surface receptor function, *Biochim. Biophys. Acta,* **415**, 81–147.
A stimulating review.
Mortimore, G. E. and Schworer, C. M., (1977), *Nature (London),* **270**, 174–176.
Variation in rates of autophagy.
Hammarstrom, S. (1975), Prostaglandins, *Ann. Rev. Biochem.,* **44**, 669–695.

Index